SIGNALS AND COMMUNICATION TECHNOLOGY

For other titles published in this series, go to
www.springer.com/series/4748

Seyed Javad Kazemitabar

Coping with Interference in Wireless Networks

Springer

Seyed Javad Kazemitabar
Dept. of EE and CS
University of California Irvine
Irvine, CA 92697-2700
USA
kazemita@hotmail.com

ISSN 1860-4862
ISBN 978-94-007-9978-3 ISBN 978-90-481-9990-7 (eBook)
DOI 10.1007/978-90-481-9990-7
Springer Dordrecht Heidelberg London New York

Softcover re-print of the Hardcover 1st edition 2011

Cover design: VTEX, Vilnius

Printed on acid-free paper

Springer is part of Springer Science+Business Media (www.springer.com)

To my parents

Preface

In this book we study the effect of interference and methods of coping with it in a wireless network. We approach the problem from three different perspectives. The first which involves physical layer is a method to cancel out interference in a multi-access channel. We consider J transmitter units each equipped with N transmit antennas over wireless Rayleigh fading channels. Previously, it was proved that when each transmitter unit has N transmit antennas, using $(J-1)N+r$ receive antennas for any $r \geq 1$, the receiver can completely separate the signals of J users. The provided diversity to each user was shown to be Nr if the units employ space-time trellis codes even if the units transmit asynchronously. Here, we consider the case when all units are synchronized and employ Quasi-Orthogonal Space-Time Block Codes ($N > 2$). It is proved that in this case a receiver with $J+r-1$ antennas, with $r \geq 1$, can separate the transmitted signals of all units and provide each unit with a diversity order of Nr.

Based on our interference cancellation technique, we then offer an array processing scheme which provides trade-off between diversity and spatial multiplexing. It is shown via simulations that this array processing scheme performs better than well-known modulation schemes, e.g. space-time block codes and BLAST, for a moderate number of receive antennas. We then derive the diversity order of these multiple antenna multi-user cancellation and detection schemes.

In our second approach we assume the physical layer did not remove interference fully. We then try to optimize our medium-access control (MAC) layer. We consider the problem of joint routing, scheduling and power control in multi-hop wireless networks. We use a linear relation between link capacity and signal to interference noise ratio in our formulation. In a previous work, using a duality approach, the optimal link scheduling and power control that minimizes the total average transmission power is found. We formulate this problem as a linear programming problem with exponential number of constraints. To cope with the exponential number of constraints, we propose an iterative algorithm based on the cutting plane method. The separation Oracle for the cutting plane algorithm turns out to be an element-wise concave optimization problem that can be effectively solved using branch and bound algorithm.

We extend the same method to find the optimal ***routing*** scheduling and power control. Simulation results show that this methodology is more efficient and scalable compared to the previously proposed algorithm.

As a third approach we investigate the connectivity of fading wireless ad-hoc networks. We first define interference, and based on that propose a few metrics of connectivity. We then study the effect of interference on connectivity based on each of those metrics.

Seyed Javad Kazemitabar

Acknowledgements

The work presented in this book was obtained through my research in the University of California Irvine. I would like to thank my advisor Professor Hamid Jafarkhani for his guidance during those years.

Contents

List of Figures

List of Tables

Chapter 1
Introduction

When two or more sources talk at the same time we will have interference. In wireless networks, interference is a common phenomenon due to sharing the medium, i.e. sky, by all the sources. Since, degrees of freedom, e.g. band-width, number of orthogonal codes, and time, are limited, interference is inevitable. It will then be useful to first, model this interference and then cope with it in different layers of network. In this book we address this issue from three different perspectives of physical layer, medium-access control layer, and network connectivity.

1 Coping with Interference in Physical Layer

Multi-user Communication and Interference Cancellation

Modern communication systems need to consider multi-user transmission (multi-access) or multi-user reception (broadcast) in their designs. During the past few decades, a large effort has been made in finding the capacity of such channels for AWGN [2–4] and Rayleigh fading [5–8] scenarios. After the introduction of notions like diversity and multiplexing [9–12] to the area of wireless communications, recently researchers have started considering them in the previously defined problems of network information theory [13, 14].

In this chapter we consider a multiple antenna multi-access scenario where by usage of receive antennas one can cancel out the interference of other users. To the best of our knowledge, the first interference cancellation technique that specifically addresses multiple antenna equipped users is [1]. In this work, Tarokh et al. consider the case where each user is equipped with N transmit antennas. They show that one can decode each user separately by using enough number of receive antennas (N times the number of users). Using properties of orthogonal space-time block codes [10, 11], Naguib et al. [15] provided an interference cancellation method that required much less number of receive antennas, i.e. as many as the number of users. However, this method was only for users equipped with 2 transmit antennas. The

S.J. Kazemitabar, *Coping with Interference in Wireless Networks*,
Signals and Communication Technology,
DOI 10.1007/978-90-481-9990-7_1,

work was improved in [16] for higher number of transmit antennas but only for $J = 2$ users. In this chapter we show why the work started by Naguib et al. cannot be applied to higher number of transmit antennas when complex constellations are used. Moreover, we offer an alternative that is based on a similar modulation scheme named Quasi-Orthogonal Space-Time Block Code (QOSTBC) [17, 18]. Our scheme also requires having receive antennas as many as the number of users only. Previously, when the number of transmit antennas, N, was more than 2, one needed approximately N times the number of users for decoding; we reduce this number to almost one Nth of what was required.

Trade off Between Rate and Diversity

Space-time codes are designed in a way to provide highest possible transmit diversity when multiple antennas are available. On the other hand, codes like BLAST [19] are designed specifically to achieve the tremendous potential rate promised by information theory [20, 21] using multiple antenna systems. The diversity order in these codes, cannot be as high as space-time codes [12]. Therefore, designing codes that provide high diversity order while maintaining rate is appreciated. Examples of such codes are golden [22] and perfect [23] codes. These two codes have full diversity and at the same time provide a high rate. The drawback of these codes, however, is that they require as many receive antennas as the number of transmit antennas, in order to use sphere decoding. The next challenge then, will be to add the property of simple decodablity and reduction in number of required receive antennas while having previous properties.

There is a direct relationship between any multi-user detection (MUD) scheme and a single user modulation scheme with multiple transmit antennas. Once the multi-user scheme is designed one can think of each user as one antenna or a group of antennas embedded in the transmit antenna array of a single transmitter unit. Using the multi-user decoding method we develop in this chapter, we provide a single user scheme that trades between diversity and spatial multiplexing. It turns out, as it is shown in the simulation results, that the new single user scheme outperforms many well-known multiple antenna schemes. We explain this part of our work in Chap. 2.

Diversity Analysis of Multi-user systems

Although, there has been a lot of work in the area of multiple antenna multi-user detection, there is a lack of performance analysis. To the best of our knowledge, a mathematical calculation of the diversity order of these MUD schemes is missing in the literature. Therefore, we were motivated to find the exact value of diversity for these schemes.

In a recent work [31] however, the authors provide a mathematical model for calculating the equivalent signal-to-noise-ratio (SNR) of different MUD methods.

Their work gives us a tool for analyzing the performance of these schemes. In this chapter we will derive the diversity order of two of the described methods based on the formulation introduced in [31]. We then extend the results to more than two transmit antenna case. We explain this part of our work in Chap. 3.

2 Coping with Interference in Medium-Access Control (MAC) Layer

Assuming the physical layer could not remove the interference fully, it will then be a problem how to design MAC. Wireless mesh networks rely on multi-hop transmission to provide connectivity between end users. In such systems it is highly desirable to minimize the total power consumption to increase energy efficiency of the system and to minimize interference with other telecommunication systems working in the same environment. Three separate set of control variables that determine the energy consumption in a network are: (i) Routing which specifies path(s) that are used to transfer data between every source-destination (S-D) pair in the network and the fraction of traffic that on average is sent through each selected path. (ii) Scheduling which specifies the set of active links that transfer data at each time slot. (iii) Power control which determines power transmission of active links at each time slot. For a given traffic matrix which specifies traffic demand between every source-destination pair our goal is to find the optimal routing, scheduling, and transmission power that minimize the total power consumption in the network.

Capacity of a link in wireless networks is not fixed and it depends on many factors including the transmission power over the link and interference caused by transmissions over other links in the network. The interdependence and coupling between link capacities is one of the fundamental characteristics that should be considered in modeling and design of wireless networks and algorithms.

The physical layer models that consider interference result in complicated mathematical relations between power and link capacity which are multi-variable and non-convex. Therefore, conventional convex optimization algorithms are not applicable anymore. Furthermore, to cope with the interdependence and coupling of variables in wireless networks, it is well-known that a cross-layer design approach that jointly optimizes routing, scheduling and power control can substantially reduce total power. In this work, we propose a cross-layer global optimal algorithm that minimizes the total power consumption in the network and consider non-convex inter-dependent model for wireless links.

Optimization approaches for wireless network resource allocation are developed in several works. Our approach is similar to [37], where the problem of joint routing, scheduling and power control for wireless multi-hop network is considered. They present a centralized algorithm to find optimal scheduling and power control to minimize the total average power consumption in the network, subject to constraints regarding the peak power and minimum data rate over each link. In addition, using the dual variable as the link cost they propose a gradient based algorithm to find the optimal routing iteratively. However, number of constraints of the scheduling and power allocation problem, which should be solved in every iteration, grows exponentially with the number of links. Hence, this approach is impractical for large size networks. We propose efficient relaxation methods and introduce an algorithm that simultaneously finds the optimal routing, scheduling and power allocation. Even though the worst case complexity of the relaxation methods can be exponential, these algorithms turn out to be very efficient for practical problems. In fact, by using the simulation results, we show that our proposed algorithms for routing, scheduling and power control is even much more efficient than the direct approach proposed in [37, 38] for scheduling and power control (with fixed routing).

Kodialam and Nandagopal [39] consider the problem of determining the achievable rates in multi-hop wireless networks. They consider a node-exclusive interference model, which means that transmitters use orthogonal channels and therefore there is no interference between transmissions of separate node pairs. They use polynomial time approximation schemes for solving the routing problem and rely on graph edge coloring methods for scheduling. In their model the transmission power over each link is fixed and hence there is no power control problem. Bhatia and Kodialam [40] consider the problem of joint routing, scheduling and power control for wireless multi-hop networks. They derive a performance guaranteed polynomial time approximation algorithm for jointly solving these three problems. Similar to [39] they also assume orthogonal channels and hence there is no interference between the channels. Lin, Lin and Shroff [42] use the same model presented in [40] and find a low complexity *distributed* algorithm for routing, scheduling and power control. Their proposed algorithm is sub-optimal and its power consumption can be twice as large as the optimal solution. Further, since their interference model is also based on the node-exclusiveness assumption, they do not consider interference caused by simultaneous transmission of separate nodes. In the simulation results section, we discuss that the adopted interference model has profound impact on the optimal routing characteristics.

Motivations and Applications

Wireless cross-layer design algorithms can be distributed or centralized. The distributed algorithms are generally designed to dynamically adapt to the network and users conditions and requirements. Therefore, they often trade-off performance for low computational complexity and communication overhead [42]. The algorithm proposed in this paper is centralized and even though it may not be applicable as an adaptive on-line algorithm it has several applications in the design and analysis of wireless networks such as:

- *Minimum Bandwidth Guarantees:* Our algorithm can be used to reserve a fraction of the time slots to efficiently provide minimum bandwidth guarantees between source-destination pairs, while for the rest of the time slots we can use a distributed algorithm to dynamically allocate the bandwidth. Distributed algorithms

may be able to provide long term (asymptotic) throughput (or fairness) guarantees, however it is much more challenging to provide short term performance guarantees with these algorithms [42, 43]. This becomes even more dramatic if we use random access methods such as 802.11 for distributed scheduling [44, 45], where starvation of some connections may occur. Under these circumstances a combination of centralized algorithms for minimum bandwidth guarantees and distributed algorithms for the rest of the capacity can be very desirable.
- *Path Selection:* Convergence of distributed routing algorithms can be very slow and before convergence we can have loops in the paths. We can use the centralized algorithm proposed in this chapter to select a good candidate set of the paths and then use distributed scheduling algorithms to distribute the load among selected paths [46]. Furthermore, our algorithm will also provide good set of the links that can be scheduled together in order to simplify the distributed scheduling.
- *Bench-marking:* Our algorithm can be used to study performance of alternative scheduling and routing algorithms and to understand how far they are from the optimal possible performance. We can also use our algorithm to detect and understand fundamental bounds and bottle-necks of a network architecture.
- *Design Insights:* Our algorithm can be used to study and answer questions regarding structure and properties of the optimal solution for wireless networks. As an example consider number of the paths that we need between every source-destination pair. In the wired networks, it is a common practice to use multiple-paths to do load-balancing in the network. However, our studies and simulations show that in many scenarios this is not a good idea in wireless networks. Note that in wireless networks, due to link coupling effect, multiple paths means greater interference which could lower the over-all capacity of the network.

Solution Overview

In [37] the problem of minimizing total power consumption of a multi-hop wireless network subject to a given offered load and a fixed routing is formulated as an element-wise concave optimization problem. Hence, the optimal power vector solutions reside on the extreme points of the feasible power region. Consequently, the optimization problem can be formulated as a linear program which has one constraint for every extreme point of the feasible power region. Therefore, complexity of this problem grows exponentially with the number of the links in the network, which makes it not scalable.

In our approach, using the relaxation method, instead of solving the LP with all constraints, at each iteration we solve a relaxed version of it with a few number of constraints. We use a non-linear optimization problem Oracle to check if the relaxed LP solution satisfies all constraints. If it does not, then the Oracle specifies one of the violated constraints that we will add to the constraint set of the relaxed LP. We continue until the relaxed LP solution satisfies all constraints. Even though in theory, it is possible that we need to add all of the constraints to the relaxed LP, practically it converges after a few iterations (relative to the number of links in the network).

We extend the Oracle to provide the routing solution too. In [37], they use a gradient based algorithm to update the routing. At each iteration, after the routing update, the optimal scheduling and power allocation should be re-computed. This is a very time consuming and inefficient procedure, so that it is reported to be efficient only for networks with at most 15 links [38]. In comparison, our simulation results show that the proposed algorithm can work for large networks with 100 links and more. Further, since we rely on global optimization algorithms, our approach can be extended and applied to other non-linear physical layer link models. We explain this part of our work in Chap. 4.

3 Effect of Interference on Connectivity

Regardless of how we deal with the interference in physical or MAC layers, it has inevitable effects on the connectivity of the network. Investigating the connectivity of radio networks goes back to four decades ago. In his pioneering work of [62], Gilbert studied the connectivity of infinite random networks relying on the so-called geometric disk model. In the geometric disk model, a random topology network is represented by a disk graph in which two nodes are directly connected if their distance is smaller than a given transmission radius. As evidenced by the works of [57, 67, 68], the connectivity of infinite random ad-hoc networks by means of the geometric disk model has recently received much attention. In addition, a survey of the literature reveals a large number of articles in the context of connectivity of ad-hoc networks with a finite number of mobile nodes. Some of the related articles in this area are [56, 58, 59, 70]. Interestingly, connectivity in random networks represented by graphs of mixed short and long edges can also be related to small world networks [74]. Although originally attractive for studying connectivity, the disk model measures connectivity relying on a pure distance-based metric which is far from the reality of wireless networks. The main disadvantages of the disk model are not considering the effects of fading, attenuation, interference, noise, and mobility.

In [63], Signal-to-Interference-Noise-Ratio (*SINR*) is proposed as the metric of connectivity in wireless ad-hoc networks. According to *SINR* metric, two nodes in a random topology are connected if their minimum *SINR* is greater than a given threshold. The two connectivity studies of [55] and [60] rely on the *SINR* model. While *SINR* is a more realistic metric of connectivity compared to the geometric disk model, it still falls short of fully capturing the connectivity phenomenon in wireless ad-hoc networks. In reality, a pair of nodes in an ad-hoc network are connected if a sequence of transmitted symbols from one can be received at another. In addition, variations of the channel in time and frequency can also affect connectivity. Utilizing Capacity (C) and/or Symbol Error Rate (*SER*) can better describe the connectivity phenomenon because those quantities are functions of not only fading, shadowing, and power but modulation and antenna configuration.

The use of space-time coding techniques in wireless networks is of special interest because it can substantially reduce the effects of multipath fading in the wireless channels through antenna diversity. Transmit antenna diversity in the form of

Space-Time Block Codes (STBCs) of [10] and [11] has been adopted in WCDMA and CDMA2000 standards. Receive antenna diversity schemes such as Maximum Ratio Combining (MRC) are already in widespread use in communication systems.

The contributions of our work are in the following areas. We introduce a pair of probabilistic connectivity metrics for wireless ad-hoc networks relying on an analysis of the time-varying fading wireless channels. We utilize central limit theorem and Gaussian approximation in our analysis to represent the combined interference and noise signal affecting the links of an ad-hoc network. Our first metric is defined based on the capacity of Multiple-Input Multiple-Output (MIMO) channels. Our second metric is defined based on the symbol error rate of such channels. We also provide a special treatment of our connectivity metrics for ergodic channels. We study this in Chap. 5.

Chapter 2
Multi-User Communication and Interference Cancellation

As the first effort to cope with interference, we try to minimize its effect in physical layer. One of the most common cases where interference will be crucial is when the base station is receiving data from several users at the same time. In this chapter we will propose a method to cancel out interference in this case using antennas at the base station.

1 The Channel Model

We model a multi-user wireless communication system where the receiving unit is equipped with $M = J + r - 1$ receive antennas, where $r \geq 1$ is the receive antenna redundancy. There are J transmitter units and each unit j, $j = 1, 2, \ldots, J$ is equipped with N antennas.

Let $c_{t,n}(j)$ denote the transmitted symbol from the n-th antenna of user j at transmission interval t and $r_{t,m}$ be the received word at the receive antenna m at the base unit. Then, for the received signal we will have

$$r_{t,m} = \sum_{j=1}^{J} \sum_{n=1}^{N} \alpha_{n,m}(j) c_{t,n}(j) + \eta_{t,m} \tag{2.1}$$

2 Interference Cancellation Using Space-Time Block Coding

It is well-known that one can separate signals sent from J different users each equipped with N transmit antennas, with $(J-1)N + 1$ receive antennas [1]. We can simply form a decoding matrix that is orthogonal to the space spanned by channel coefficients of the users to be eliminated:

$$\mathbf{R}_t = \mathbf{C}_t \mathbf{H} + \mathcal{N}_t \tag{2.2}$$

S.J. Kazemitabar, *Coping with Interference in Wireless Networks*,
Signals and Communication Technology,
DOI 10.1007/978-90-481-9990-7_2,

where

$$\begin{aligned}
\mathbf{C}_t &= (\mathbf{C}_t(1), \mathbf{C}_t(2), \ldots, \mathbf{C}_t(J)) \\
\mathbf{R}_t &= (r_{t,1}, r_{t,2}, \ldots, r_{t,M}) \\
\mathcal{N}_t &= (\eta_{t,1}, \eta_{t,2}, \ldots, \eta_{t,M}) \\
\mathbf{H} &= \left(\mathbf{H}(1)^T | \mathbf{H}(2)^T | \cdots | \mathbf{H}(J)^T\right)
\end{aligned} \tag{2.3}$$

with

$$\mathbf{C}_t(j) = (c_{t,1}(j), c_{t,2}(j), \ldots, c_{t,N}(j))$$

$$\mathbf{H}(j) = \begin{pmatrix} \alpha_{1,1}(j) & \alpha_{1,2}(j) & \cdots & \alpha_{1,M}(j) \\ \alpha_{2,1}(j) & \alpha_{2,2}(j) & \cdots & \alpha_{2,M}(j) \\ \vdots & \vdots & \ddots & \vdots \\ \alpha_{N,1}(j) & \alpha_{N,2}(j) & \cdots & \alpha_{N,M}(j) \end{pmatrix} \tag{2.4}$$

Therefore, one can rewrite (3.4) as follows:

$$\mathbf{R}_t = \sum_{j=1}^{J} \mathbf{C}_t(j)\mathbf{H}(j) + \mathcal{N}_t \tag{2.5}$$

To decode user 1, one can simply find a zero-forcing (ZF) matrix $\mathbf{Z}$ such that

$$\begin{aligned}
&\mathbf{H}(1)\mathbf{Z} \neq \mathbf{0} \\
&\mathbf{H}(j)\mathbf{Z} = \mathbf{0} \quad \text{for } j \neq 1
\end{aligned} \tag{2.6}$$

In other words, $\mathbf{Z}$ should null the space spanned by the row vectors of all $\mathbf{H}(j)$s, for $j = 2, 3, \ldots, J$. Also, it should not null at least one row vector of $\mathbf{H}(1)$. Since all the rows of $\mathbf{H}(j)$s might be linearly independent, the dimension of $\mathbf{Z}$, i.e. M, must be at least equal to the number of these rows, or $(J-1)N+1$. Each antenna group (user) can employ a modulation scheme to benefit transmit diversity; as if it is the only group that is sending data.

One might naturally think that using some smart coding in space and/or time, may reduce the number of required receive antennas. In fact that is true; in [15] it is shown that when $N = 2$ and users are equipped with Alamouti code, i.e. Orthogonal Space-Time Block Code (OSTBC) for $N = 2$, the number of required receive antennas is reduced to J. Also, in [16], same task is accomplished for a larger group of OSTBCs when only $J = 2$ users exist. To the best of our knowledge, for users with more than 2 transmit antennas and for general number of users, such an example does not exist in the literature. In what follows we describe the algorithm in [15] except that we use ZF instead of minimum mean square error (MMSE) that requires matrix inversion. We then show why it is not possible to repeat what Naguib et al. did for higher order OSTBCs for an arbitrary number of users[1] and then offer an alternative.

Consider 2 users each transmitting an Alamouti block to a receiver unit equipped with at least 2 receive antennas. The received signal can be written in the following format:

$$\begin{pmatrix} r_{1,m} \\ r_{2,m} \end{pmatrix} = \begin{pmatrix} s_1(1) & s_2(1) \\ -s_2^*(1) & s_1^*(1) \end{pmatrix} \cdot \begin{pmatrix} \alpha_{1,m}(1) \\ \alpha_{2,m}(1) \end{pmatrix} + \begin{pmatrix} s_1(2) & s_2(2) \\ -s_2^*(2) & s_1^*(2) \end{pmatrix} \cdot \begin{pmatrix} \alpha_{1,m}(2) \\ \alpha_{2,m}(2) \end{pmatrix} + \begin{pmatrix} \eta_{1,m} \\ \eta_{2,m} \end{pmatrix} \tag{2.7}$$

The idea behind interference cancellation arises from separate decodability of each symbol; at each receive antenna we perform the decoding algorithm as if there is only one user. This user will be the one the effect of whom we want to cancel out. Then, we simply subtract the soft-decoded value of each symbol in one of the receive antennas from the rest and as a result remove the effect of that user. This procedure is presented in the following lines:

$$\begin{pmatrix} r_{1,m} \\ r_{2,m}^* \end{pmatrix} = \begin{pmatrix} \alpha_{1,m}(1) & \alpha_{2,m}(1) \\ \alpha_{2,m}^*(1) & -\alpha_{1,m}^*(1) \end{pmatrix} \cdot \begin{pmatrix} s_1(1) \\ s_2(1) \end{pmatrix} + \begin{pmatrix} \alpha_{1,m}(2) & \alpha_{2,m}(2) \\ \alpha_{2,m}^*(2) & -\alpha_{1,m}^*(2) \end{pmatrix} \cdot \begin{pmatrix} s_1(2) \\ s_2(2) \end{pmatrix} + \begin{pmatrix} \eta_{1,m} \\ \eta_{2,m}^* \end{pmatrix} \tag{2.8}$$

As one can easily check, the matrix multiplied to $(s_1(1), s_2(1))^T$ is multiple of a unitary matrix. Therefore, we can simply write

$$\begin{pmatrix} \alpha_{1,m}^*(1) & \alpha_{2,m}(1) \\ \alpha_{2,m}^*(1) & -\alpha_{1,m}(1) \end{pmatrix} \begin{pmatrix} r_{1,m} \\ r_{2,m}^* \end{pmatrix} = (|\alpha_{1,m}(1)|^2 + |\alpha_{2,m}(1)|^2) \begin{pmatrix} s_1(1) \\ s_2(1) \end{pmatrix}$$
$$+ \begin{pmatrix} \alpha_{1,m}^*(1) & \alpha_{2,m}(1) \\ \alpha_{2,m}^*(1) & -\alpha_{1,m}(1) \end{pmatrix} \begin{pmatrix} \alpha_{1,m}(2) & \alpha_{2,m}(2) \\ \alpha_{2,m}^*(2) & -\alpha_{1,m}^*(2) \end{pmatrix} \begin{pmatrix} s_1(2) \\ s_2(2) \end{pmatrix} + \begin{pmatrix} \eta'_{1,m} \\ \eta'_{2,m} \end{pmatrix} \tag{2.9}$$

for $m = 1, 2, \ldots, M$. $\eta'_{i,m}$ is a zero mean Gaussian random variable, with variance $\frac{2(|\alpha_{1,m}(1)|^2+|\alpha_{2,m}(1)|^2)}{SNR}$. The above analysis is a part of decoding for any orthogonal space-time block code [25, Chap. 4]. To completely eliminate the effect of user No. 1, it remains to divide (9) by $(|\alpha_{1,m}(1)|^2 + |\alpha_{2,m}(1)|^2)$ and subtract the terms

[1] When there are only 2 users, it is possible to do interference cancellation with a class of OSTBCs [16].

for $m = 1$ from that of $m = 2, \dots, M$. The resulting terms will be

$$\left\{ \frac{1}{|\alpha_{1,m}(1)|^2 + |\alpha_{2,m}(1)|^2} \begin{pmatrix} \alpha_{1,m}^*(1) & \alpha_{2,m}(1) \\ \alpha_{2,m}^*(1) & -\alpha_{1,m}(1) \end{pmatrix} \begin{pmatrix} \alpha_{1,m}(2) & \alpha_{2,m}(2) \\ \alpha_{2,m}(2)^* & -\alpha_{1,m}(2)^* \end{pmatrix} \right.$$
$$\left. - \frac{1}{|\alpha_{1,1}(1)|^2 + |\alpha_{2,1}(1)|^2} \begin{pmatrix} \alpha_{1,1}^*(1) & \alpha_{2,1}(1) \\ \alpha_{2,1}^*(1) & -\alpha_{1,1}(1) \end{pmatrix} \begin{pmatrix} \alpha_{1,1}(2) & \alpha_{2,1}(2) \\ \alpha_{2,1}^*(2) & -\alpha_{1,1}^*(2) \end{pmatrix} \right\}$$
$$\times \begin{pmatrix} s_1(2) \\ s_2(2) \end{pmatrix} + \begin{pmatrix} \eta_{1,m}'' \\ \eta_{2,m}'' \end{pmatrix} \tag{2.10}$$

for $m = 2, \dots, M$. The distribution of $\eta_{i,m}''$ is Gaussian, and its variance is equal to $\frac{2}{SNR(|\alpha_{1,m}(1)|^2+|\alpha_{2,m}(1)|^2)} + \frac{2}{SNR(|\alpha_{1,1}(1)|^2+|\alpha_{2,1}(1)|^2)}$. It can be easily shown that the overall matrix multiplied by $(s_1(2), s_2(2))^T$ above is also a multiple of a unitary matrix. Therefore, the system in (2.10) is equivalent to one that transmits an orthogonal design. We can simply apply Maximum-Likelihood (ML) decoding, which is separate for each of the symbols due to having a unitary channel matrix. Note that only $M = J = 2$ receive antennas are needed for gaining full transmit diversity, $N = 2$; using more receive antennas will result in a multiplicative receive diversity, such that the overall diversity will be $2r$, with $r = M - J + 1$.

Lemma 1 *The interference cancellation technique performed on Alamouti structure cannot be extended to higher order OSTBCs when having more than 2 users.*

Proof After eliminating the first user there will be K equations left each produced by eliminating one of the symbols $(s_1(1), s_2(1), \dots, s_K(1))^T$, where K is the number of symbols in the block code. The algorithm described above requires that after elimination of the first user, the resulting equations be in the following format

$$\sum_{i=2}^{J} \mathbf{U_i} \cdot \underline{\mathbf{s_i}} \tag{2.11}$$

where $\underline{\mathbf{s_i}} = (s_1(i), s_2(i), \dots, s_K(i))^T$ and $\mathbf{U_i}$ is a multiple of a unitary matrix. A set of equations with the form of $\mathbf{U} \cdot \underline{\mathbf{s}}$ is equivalent to usage of a $K \times N$ orthogonal block code in the transmitter. A block code sending K symbols in K time slots is rate one. However, we know that rate one complex orthogonal design is impossible for more than 2 transmit antennas [11, 24]. Therefore, after eliminating the first user, the resulting system cannot enjoy simple decoding as it did for the case of 2 transmit antennas in [15]. Therefore, users No. 2 and above cannot be cancelled out. This means that the algorithm used in [15] cannot be extended to more than 2 transmit antennas for an arbitrary number of users. □

Note that all this discussion happens when the first user is cancelled. Therefore, we might be able to perform interference cancellationinterference cancellation using

OSTBCs for the case of $J = 2$ as [16] suggests. In addition, since there exist rate one *real* orthogonal designs, we can apply the same algorithm to them as we show in the following lemmas.

Definition A generalized real orthogonal design is a $T \times N$ matrix $\mathbf{C}$ with real entries $s_1, -s_1, s_2, -s_2, \ldots, s_K, -s_K$ such that

$$\mathbf{C}^T\mathbf{C} = \kappa\left(s_1^2 + s_2^2 + \cdots + s_K^2\right)I_N \tag{2.12}$$

Lemma 2 *A generalized real orthogonal design can be written as*

$$\mathbf{C} = \sum_{k=1}^{K} s_k \mathbf{E}_k \tag{2.13}$$

where matrices $\mathbf{E}_k$ are $T \times N$ real matrices and satisfy

$$\begin{cases} \mathbf{E}_k^T\mathbf{E}_{k'} + \mathbf{E}_{k'}^T\mathbf{E}_k = 0_N & k \neq k' \\ \mathbf{E}_k^T\mathbf{E}_k = I_N & k = 1, 2, \ldots, K \end{cases} \tag{2.14}$$

Proof Refer to [25]. □

Lemma 3 *For any number of transmit antennas, with usage of real orthogonal space-time block codes, and using a real constellation, we can simply decode a group of J users given the same number of receive antennas.*

Proof Consider a real channel with channel matrix $\mathbf{H}$ and noise vector $\mathcal{N}$. Assume we have J users each equipped with N antennas and transmitting a $T \times N$ real orthogonal code $\mathbf{C}$. At each receive antenna m, for $m = 1, 2, \ldots, M$ we have

$$\begin{aligned} \mathbf{R}_m^T &= \sum_{j=1}^{J} \mathbf{H}_m^T(j) \cdot \mathbf{C}^T(j) + \mathcal{N}_m^T = \sum_{j=1}^{J} \mathbf{H}_m^T(j) \sum_{k=1}^{K} s_k(j)\mathbf{E}_k^T + \mathcal{N}_m^T \\ &= \sum_{j=1}^{J}\sum_{k=1}^{K} s_k(j)\boldsymbol{\Omega}_k^m(j) + \mathcal{N}_m^T \\ &= \sum_{j=1}^{J} (s_1(j), s_2(j), \ldots, s_K(j)) \cdot \boldsymbol{\Omega}_m(j) + \mathcal{N}_m^T \end{aligned} \tag{2.15}$$

where $\boldsymbol{\Omega}_k^m(j) = \mathbf{H}_m^T(j)\mathbf{E}_k^T$ is the kth row of a $K \times T$ matrix $\boldsymbol{\Omega}_m(j)$ and contains T elements. These elements are a function of N path gains $\alpha_{1m}(j), \alpha_{2m}(j), \ldots,$ $\alpha_{Nm}(j)$. It can be easily shown that [25]

$$\boldsymbol{\Omega}_m(j) \cdot \boldsymbol{\Omega}_m^T(j) = \left(\sum_{n=1}^{N} \alpha_{nm}^2(j)\right) I_K \tag{2.16}$$

For example, let us assume we want to eliminate the effect of user No. 1. We simply multiply each $\mathbf{R}_m$ by $\boldsymbol{\Omega}_m^T(1)$, divide it by $\sum_{n=1}^{N} \alpha_{nm}^2(1)$ and subtract the term for $m = 1$ from that of $m = 2, \ldots, M$:

$$\frac{\mathbf{R}_m^T \cdot \boldsymbol{\Omega}_m^T(1)}{\sum_{n=1}^{N} \alpha_{nm}^2(1)} - \frac{\mathbf{R}_1^T \cdot \boldsymbol{\Omega}_1^T(1)}{\sum_{n=1}^{N} \alpha_{n1}^2(1)} = \sum_{j=2}^{J} (s_1(j), s_2(j), \ldots, s_K(j)) \cdot \left\{ \frac{\boldsymbol{\Omega}_m(j) \cdot \boldsymbol{\Omega}_m^T(1)}{\sum_{n=1}^{N} \alpha_{nm}^2(1)} - \frac{\boldsymbol{\Omega}_1(j) \cdot \boldsymbol{\Omega}_1^T(1)}{\sum_{n=1}^{N} \alpha_{n1}^2(1)} \right\} + \mathcal{N}' \tag{2.17}$$

To prove that we can continue eliminating other users, we need to show that the following matrix is also multiple of a unitary matrix:

$$\frac{\boldsymbol{\Omega}_m(j) \cdot \boldsymbol{\Omega}_m^T(1)}{\sum_{n=1}^{N} \alpha_{nm}^2(1)} - \frac{\boldsymbol{\Omega}_1(j) \cdot \boldsymbol{\Omega}_1^T(1)}{\sum_{n=1}^{N} \alpha_{n1}^2(1)} \tag{2.18}$$

For that, it would be enough to show $\boldsymbol{\Omega}_{m_1}(j_1) \cdot \boldsymbol{\Omega}_{m_2}^T(j_2)$ is a multiple of a unitary matrix.[2] The latter is true due to the following:

$$\begin{aligned} &\left[\boldsymbol{\Omega}_{m_1}(j_1) \cdot \boldsymbol{\Omega}_{m_2}^T(j_2)\right]^T \left[\boldsymbol{\Omega}_{m_1}(j_1) \cdot \boldsymbol{\Omega}_{m_2}^T(j_2)\right] \\ &= \boldsymbol{\Omega}_{m_2}(j_2) \cdot \boldsymbol{\Omega}_{m_1}^T(j_1) \cdot \boldsymbol{\Omega}_{m_1}(j_1) \cdot \boldsymbol{\Omega}_{m_2}^T(j_2) \end{aligned} \tag{2.19}$$

Since $K = T$ [25], $\boldsymbol{\Omega}_m$s are $K \times K$ square matrices and (2.16) implies

$$\boldsymbol{\Omega}_m^T(j) \cdot \boldsymbol{\Omega}_m(j) = \left(\sum_{n=1}^{N} \alpha_{nm}^2(j)\right) I_K \tag{2.20}$$

Therefore, (2.19) can be simplified as following

$$\boldsymbol{\Omega}_{m_2}(j_2) \cdot \left(\sum_{n=1}^{N} \alpha_{nm_1}^2(j_1)\right) I_K \cdot \boldsymbol{\Omega}_{m_2}^T(j_2) = \left(\sum_{n=1}^{N} \alpha_{nm_1}^2(j_1)\right)\left(\sum_{n=1}^{N} \alpha_{nm_2}^2(j_2)\right) I_K \tag{2.21}$$

This concludes the proof that the matrix in (2.18) is a multiple of a unitary matrix. Therefore, once we eliminate the first user we can apply the same method on the remaining $M - 1$ signals and eliminate the second user. The reason is because the corresponding Ω matrix at each stage of elimination is square and multiple of a unitary matrix and the above proof guarantees the same property for the matrix of the next stage. □

[2] Since we know that the two subtracted terms will have the same format and will keep it after subtraction.

However, as discussed earlier, *complex* OSTBCs, except for $N = 2$, are not good candidates for interference cancellation when the goal is reducing the number of receive antennas. In next section, we offer another modulation scheme that fulfills the task for higher number of transmit antennas.

3 Interference Cancellation Using Quasi-Orthogonal Space-Time Block Coding

We use *Quasi-Orthogonal Space-Time Block Codes (QOSTBCs)* as follows [17]:

$$\mathbf{C} = \begin{pmatrix} s_1 & s_2 & s_3 & s_4 \\ -s_2^* & s_1^* & -s_4^* & s_3^* \\ s_3 & s_4 & s_1 & s_2 \\ -s_4^* & s_3^* & -s_2^* & s_1^* \end{pmatrix} \tag{2.22}$$

In this design each column of the generator matrix is orthogonal to all other columns except one. As a result a pairwise decoding of the symbols is possible. Using rotation for some of the symbols, full diversity QOSTBCs are realizable [18, 23].

We describe the decoding when there is one user. The multi-user case is next to be studied. Assuming perfect channel state information is available, the receiver computes the decision metric

$$\sum_{m=1}^{M}\sum_{t=1}^{4}\left| r_{t,m} - \sum_{n=1}^{4} \alpha_{n,m}(1) c_{t,n}(1) \right|^2 \tag{2.23}$$

over all possible symbols to replace $s_1, \dots, s_4$ in $\mathbf{C}$ and decides in favor of constellation symbols that minimize this sum. Since we have only one user and for simplicity specify one receive antenna, we do not mention indexing of group or receive antenna in the rest of this section. Simple algebraic manipulation shows that ML decoding for the code in (2.22) is equivalent to minimizing the following sum [17]:

$$f_{13}(s_1, s_3) + f_{24}(s_2, s_4) \tag{2.24}$$

where $f_{13}(s_1, s_3)$ is independent of (s_2, s_4) and $f_{24}(s_2, s_4)$ is independent of (s_1, s_3). Therefore, the pairs (s_1, s_3) and (s_2, s_4) can be decoded separately and the scheme is pairwise decodable.

In [27] a new decoding method was introduced by which differential decoding of a QOSTBC for non-coherent systems became possible. Let us review this decoding method. For $M = 1$ receive antenna, let us define the received signals at four time slots by r_1, r_2, r_3, r_4. Then, the set of input-output equations is equivalent to [27]:

$$\mathbf{R}_1 = \mathbf{S}_1 \mathbf{H}_1 + \mathcal{N}_1 \tag{2.25}$$

where

$$\mathbf{R}_1 = (r_{13,1}, r_{24,1})^T = (r_1 + r_3, r_2 + r_4)^T$$
$$\mathbf{H}_1 = (\alpha_{13,1}, \alpha_{24,1})^T = (\alpha_1 + \alpha_3, \alpha_2 + \alpha_4)^T$$
$$\mathcal{N}_1 = (\eta_1 + \eta_3, \eta_2 + \eta_4)^T$$
$$\mathbf{S}_1 = \begin{pmatrix} s_{13,1} & s_{24,1} \\ -s_{24,1}^* & s_{13,1}^* \end{pmatrix} = \begin{pmatrix} s_1 + s_3 & s_2 + s_4 \\ -(s_2^* + s_4^*) & s_1^* + s_3^* \end{pmatrix}$$

and

$$\mathbf{R}_2 = \mathbf{S}_2\mathbf{H}_2 + \mathcal{N}_2 \tag{2.26}$$

where

$$\mathbf{R}_2 = (r_{13,2}, r_{24,2})^T = (r_1 - r_3, r_2 - r_4)^T$$
$$\mathbf{H}_2 = (\alpha_{13,2}, \alpha_{24,2})^T = (\alpha_1 - \alpha_3, \alpha_2 - \alpha_4)^T$$
$$\mathcal{N}_2 = (\eta_1 - \eta_3, \eta_2 - \eta_4)^T$$
$$\mathbf{S}_2 = \begin{pmatrix} s_{13,2} & s_{24,2} \\ -s_{24,2}^* & s_{13,2}^* \end{pmatrix} = \begin{pmatrix} s_1 - s_3 & s_2 - s_4 \\ -(s_2^* - s_4^*) & s_1^* - s_3^* \end{pmatrix}$$

We can consider (2.25) and (2.26) as two equivalent subsystems, each of which has two transmit antennas. We can see that $\mathbf{S}_1$ and $\mathbf{S}_2$ have the structure of Alamouti code [10]. This is the key property in multi-user decoding as we discuss in the next section. The ML decoding metric for this system is[3]

$$\begin{aligned} X = &|r_{13,1} - \alpha_{13,1}s_{13,1} - \alpha_{24,1}s_{24,1}|^2 + |r_{24,1} + \alpha_{13,1}s_{24,1}^* + \alpha_{24,1}s_{13,1}|^2 \\ &+ |r_{13,2} - \alpha_{13,2}s_{13,2} - \alpha_{24,2}s_{24,2}|^2 \\ &+ |r_{24,2} + \alpha_{13,2}s_{24,2}^* - \alpha_{24,2}s_{13,2}^*|^2 \end{aligned} \tag{2.27}$$

We need to find symbols s_1, s_2, s_3, s_4 that minimize X. Expanding the expression for X we get

$$X = |r_1|^2 + |r_2|^2 + |r_3|^2 + |r_4|^2 + 2f_{13}(s_1, s_3) + 2f_{24}(s_2, s_4) \tag{2.28}$$

Therefore, the choice of $\{s_1, s_2, s_3, s_4\}$ that minimize X, will minimize the ML metric of the original system as well. In other words, our transformation is lossless, and the error performance is the same as the optimal decoding of the system in (2.22). Then, using the method in Sect. III for multi-user detection of Alamouti equipped transmitters, we decode the whole group.

[3]The reason we can write ML metric like this is because the new noise terms are still independent and Gaussian.

Recalling the model in Sect. II and having in mind that users are sending QOSTBCs, at each receive antenna m we receive the following signals in the four time slots

$$\begin{aligned}
r_{1,m} &= \sum_{j=1}^{J} \alpha_{1,m}(j)s_1(j) + \alpha_{2,m}(j)s_2(j) + \alpha_{3,m}(j)s_3(j) + \alpha_{4,m}(j)s_4(j) + \eta_{1,m} \\
r_{2,m} &= \sum_{j=1}^{J} -\alpha_{1,m}(j)s_2^*(j) + \alpha_{2,m}(j)s_1^*(j) - \alpha_{3,m}(j)s_4^*(j) + \alpha_{4,m}(j)s_3^*(j) + \eta_{2,m} \\
r_{3,m} &= \sum_{j=1}^{J} \alpha_{1,m}(j)s_3(j) + \alpha_{2,m}(j)s_4(j) + \alpha_{3,m}(j)s_1(j) + \alpha_{4,m}(j)s_2(j) + \eta_{3,m} \\
r_{4,m} &= \sum_{j=1}^{J} -\alpha_{1,m}(j)s_4^*(j) + \alpha_{2,m}(j)s_3^*(j) - \alpha_{3,m}(j)s_2^*(j) + \alpha_{4,m}(j)s_1^*(j) + \eta_{4,m}
\end{aligned} \tag{2.29}$$

where $\eta_{i,m}$s are i.i.d. zero mean Gaussian random variables with variance $\frac{4}{SNR}$. Similarly, for the signals at each receive antenna we form two equivalent 2-antenna systems as follows

$$\begin{aligned}
\mathbf{R}_{1,m} &= \begin{pmatrix} r_{1,m} + r_{3,m} \\ r_{2,m} + r_{4,m} \end{pmatrix} = \sum_{j=1}^{J} \begin{pmatrix} s_1(j) + s_3(j) & s_2(j) + s_4(j) \\ -(s_2(j) + s_4(j))^* & (s_1(j) + s_3(j))^* \end{pmatrix} \\
&\quad \times \begin{pmatrix} \alpha_{1,m}(1) + \alpha_{3,m}(1) \\ \alpha_{2,m}(1) + \alpha_{4,m}(1) \end{pmatrix} + \begin{pmatrix} \eta_{1,m} + \eta_{3,m} \\ \eta_{2,m} + \eta_{4,m} \end{pmatrix} \\
\mathbf{R}_{2,m} &= \begin{pmatrix} r_{1,m} - r_{3,m} \\ r_{2,m} - r_{4,m} \end{pmatrix} = \sum_{j=1}^{J} \begin{pmatrix} s_1(j) - s_3(j) & s_2(j) - s_4(j) \\ -(s_2(j) - s_4(j))^* & (s_1(j) - s_3(j))^* \end{pmatrix} \\
&\quad \times \begin{pmatrix} \alpha_{1,m}(1) - \alpha_{3,m}(1) \\ \alpha_{2,m}(1) - \alpha_{4,m}(1) \end{pmatrix} + \begin{pmatrix} \eta_{1,m} - \eta_{3,m} \\ \eta_{2,m} - \eta_{4,m} \end{pmatrix}
\end{aligned} \tag{2.30}$$

Without loss of generality we assume that we eliminate the effect of user No. 1 first. By applying complex conjugation on the second row of both systems we get

$$\begin{aligned}
\mathbf{R'}^{T}_{1,m} &= \begin{pmatrix} r_{1,m} + r_{3,m} \\ r_{2,m}^* + r_{4,m}^* \end{pmatrix}^T = \sum_{j=1}^{J} (s_1(j) + s_3(j), s_2(j) + s_4(j)) \\
&\cdot \begin{pmatrix} \alpha_{1,m}(j) + \alpha_{3,m}(j) & (\alpha_{2,m}(j) + \alpha_{4,m}(j))^* \\ \alpha_{2,m}(j) + \alpha_{4,m}(j) & -(\alpha_{1,m}(j) + \alpha_{3,m}(j))^* \end{pmatrix} + \begin{pmatrix} \eta_{1,m} + \eta_{3,m} \\ \eta_{2,m}^* + \eta_{4,m}^* \end{pmatrix}^T
\end{aligned}$$

$$= \sum_{j=1}^{J} (s_1(j) + s_3(j), s_2(j) + s_4(j)) \cdot \mathcal{G}(\alpha_{1,m}(j) + \alpha_{3,m}(j), \alpha_{2,m}(j) + \alpha_{4,m}(j)) + \mathcal{N}_{1,m} \tag{2.31}$$

and

$$\mathbf{R}'^{T}_{2,m} = \begin{pmatrix} r_{1,m} - r_{3,m} \\ r^*_{2,m} - r^*_{4,m} \end{pmatrix}^T = \sum_{j=1}^{J} (s_1(j) - s_3(j), s_2(j) - s_4(j)) \cdot \begin{pmatrix} \alpha_{1,m}(j) - \alpha_{3,m}(j) & (\alpha_{2,m}(j) - \alpha_{4,m}(j))^* \\ \alpha_{2,m}(j) - \alpha_{4,m}(j) & -(\alpha_{1,m}(j) - \alpha_{3,m}(j))^* \end{pmatrix} + \begin{pmatrix} \eta_{1,m} - \eta_{3,m} \\ \eta^*_{2,m} - \eta^*_{4,m} \end{pmatrix}^T$$
$$= \sum_{j=1}^{J} (s_1(j) - s_3(j), s_2(j) - s_4(j)) \cdot \mathcal{G}(\alpha_{1,m}(j) - \alpha_{3,m}(j), \alpha_{2,m}(j) - \alpha_{4,m}(j)) + \mathcal{N}_{2,m} \tag{2.32}$$

where

$$\mathcal{G}(x, y) = \begin{pmatrix} x & y^* \\ y & -x^* \end{pmatrix} \tag{2.33}$$

The advantage of representing the received signals in the above format is hidden in the structure of the equivalent channel matrices:

$$\mathbf{\Omega}_{1,m}(j) = \begin{pmatrix} \alpha_{1,m}(j) + \alpha_{3,m}(j) & (\alpha_{2,m}(j) + \alpha_{4,m}(j))^* \\ \alpha_{1,m}(j) + \alpha_{3,m}(j) & -(\alpha_{1,m}(j) + \alpha_{3,m}(j))^* \end{pmatrix}$$
$$= \mathcal{G}(\alpha_{1,m}(j) + \alpha_{3,m}(j), \alpha_{2,m}(j) + \alpha_{4,m}(j))$$
$$\mathbf{\Omega}_{2,m}(j) = \begin{pmatrix} \alpha_{1,m}(j) - \alpha_{3,m}(j) & (\alpha_{2,m}(j) - \alpha_{4,m}(j))^* \\ \alpha_{1,m}(j) - \alpha_{3,m}(j) & -(\alpha_{1,m}(j) - \alpha_{3,m}(j))^* \end{pmatrix}$$
$$= \mathcal{G}(\alpha_{1,m}(j) - \alpha_{3,m}(j), \alpha_{2,m}(j) - \alpha_{4,m}(j))$$

Since both $\mathbf{\Omega}_{1,m}(j)$ and $\mathbf{\Omega}_{2,m}(j)$ are multiples of a unitary matrix, we can simply separate and therefore eliminate the effect of each group from the received signal.

$$\mathbf{R}'^{T}_{1,m} \cdot \mathbf{\Omega}^{\dagger}_{1,m}(1)$$
$$= (|\alpha_{1,m}(1) + \alpha_{3,m}(1)|^2 + |\alpha_{2,m}(1) + \alpha_{4,m}(1)|^2)(s_1(j) + s_3(1), s_2(1) + s_4(1))$$
$$+ \sum_{j=2}^{J} (s_1(j) + s_3(j), s_2(j) + s_4(j)) \cdot \mathcal{G}'((\alpha_{1,m}(j) + \alpha_{3,m}(j))$$

$$\times (\alpha_{1,m}(j) + \alpha_{3,m}(j))^* + (\alpha_{2,m}(j) + \alpha_{4,m}(j))^*(\alpha_{2,m}(1) + \alpha_{4,m}(1))$$
$$\times (\alpha_{1,m}(j) + \alpha_{3,m}(j))(\alpha_{2,m}(j) + \alpha_{4,m}(j))^*$$
$$- (\alpha_{2,m}(j) + \alpha_{4,m}(j))^*(\alpha_{1,m}(1) + \alpha_{3,m}(1))) + \mathcal{N}'_{1,m} \tag{2.34}$$

$$\mathbf{R'}^T_{2,m} \cdot \mathbf{\Omega}^\dagger_{2,m}(1)$$
$$= (|\alpha_{1,m}(1) - \alpha_{3,m}(1)|^2 + |\alpha_{2,m}(1) - \alpha_{4,m}(1)|^2)(s_1(j) - s_3(1), s_2(1) - s_4(1))$$
$$+ \sum_{j=2}^{J} (s_1(j) - s_3(j), s_2(j) - s_4(j)) \cdot \mathcal{G}'((\alpha_{1,m}(j) - \alpha_{3,m}(j))$$
$$\times (\alpha_{1,m}(j) - \alpha_{3,m}(j))^* + (\alpha_{2,m}(j) - \alpha_{4,m}(j))^*(\alpha_{2,m}(1) - \alpha_{4,m}(1))$$
$$\times (\alpha_{1,m}(j) - \alpha_{3,m}(j))(\alpha_{2,m}(j) - \alpha_{4,m}(j))^*$$
$$- (\alpha_{2,m}(j) - \alpha_{4,m}(j))^*(\alpha_{1,m}(1) - \alpha_{3,m}(1))) + \mathcal{N}'_{2,m} \tag{2.35}$$

where

$$\mathcal{G}'(x, y) = \begin{pmatrix} x & y \\ -y^* & x^* \end{pmatrix} \tag{2.36}$$

$\mathcal{N}'_{1,m}$ and $\mathcal{N}'_{2,m}$ are the new noise vectors corresponding to the mth receive antenna. Noting that $\mathbf{\Omega}_{i,m}$ matrices are multiples of unitary, the distribution of each element of $\mathcal{N}'_{1,m}$ will be Gaussian with variance $\frac{8(|\alpha_{1,m}(1)+\alpha_{3,m}(1)|^2+|\alpha_{2,m}(1)+\alpha_{4,m}(1)|^2)}{SNR}$. The two elements will still be i.i.d. Same thing is true for $\mathcal{N}'_{2,m}$ with a change of sign.

Let us reconsider (2.34) and (2.35) for all receive antennas $m = 1, 2, \ldots, J + r - 1$. If we subtract the expression for the first receive antenna, $m = 1$, from that of the other antennas we will have $J + r - 2$ equations as follows

$$\frac{\mathbf{R'}^T_{1,m} \cdot \mathbf{\Omega}^\dagger_{1,m}(1)}{|\alpha_{1,m}(1) + \alpha_{3,m}(1)|^2 + |\alpha_{2,m}(1) + \alpha_{4,m}(1)|^2}$$
$$- \frac{\mathbf{R'}^T_{1,1} \cdot \mathbf{\Omega}^\dagger_{1,1}(1)}{|\alpha_{1,1}(1) + \alpha_{3,1}(1)|^2 + |\alpha_{2,1}(1) + \alpha_{4,1}(1)|^2}$$
$$= \sum_{j=2}^{J} (s_1(j) + s_3(j), s_2(j) + s_4(j)) \cdot \mathcal{G}'(A_{1m}(j), B_{1m}(j))$$
$$+ \frac{\mathcal{N}'_{1,m}}{|\alpha_{1,m}(1) + \alpha_{3,m}(1)|^2 + |\alpha_{2,m}(1) + \alpha_{4,m}(1)|^2}$$
$$- \frac{\mathcal{N}'_{1,1}}{|\alpha_{1,1}(1) + \alpha_{3,1}(1)|^2 + |\alpha_{2,1}(1) + \alpha_{4,1}(1)|^2} \tag{2.37}$$

and

$$\frac{\mathbf{R}'^{T}_{2,m} \cdot \mathbf{\Omega}^{\dagger}_{2,m}(1)}{|\alpha_{1,m}(1) - \alpha_{3,m}(1)|^2 + |\alpha_{2,m}(1) - \alpha_{4,m}(1)|^2} - \frac{\mathbf{R}'^{T}_{2,1} \cdot \mathbf{\Omega}^{\dagger}_{2,1}(1)}{|\alpha_{1,1}(1) - \alpha_{3,1}(1)|^2 + |\alpha_{2,1}(1) - \alpha_{4,1}(1)|^2}$$
$$= \sum_{j=2}^{J} (s_1(j) - s_3(j), s_2(j) - s_4(j)) \cdot \mathcal{G}'(A_{2m}(j), B_{2m}(j))$$
$$+ \frac{\mathcal{N}'_{2,m}}{|\alpha_{1,m}(1) - \alpha_{3,m}(1)|^2 + |\alpha_{2,m}(1) - \alpha_{4,m}(1)|^2} - \frac{\mathcal{N}'_{2,1}}{|\alpha_{1,1}(1) - \alpha_{3,1}(1)|^2 + |\alpha_{2,1}(1) - \alpha_{4,1}(1)|^2} \tag{2.38}$$

where

$$A_{1m}(j) = \frac{(\alpha_{1,m}(j) + \alpha_{3,m}(j))(\alpha_{1,m}(1) + \alpha_{3,m}(1))^* + (\alpha_{2,m}(j) + \alpha_{4,m}(j))^*(\alpha_{2,m}(j) + \alpha_{4,m}(j))}{|\alpha_{1,m}(1) + \alpha_{3,m}(1)|^2 + |\alpha_{2,m}(1) + \alpha_{4,m}(1)|^2}$$
$$- \frac{(\alpha_{1,1}(j) + \alpha_{3,1}(j))(\alpha_{1,1}(1) + \alpha_{3,1}(1))^* + (\alpha_{2,1}(j) + \alpha_{4,1}(j))^*(\alpha_{2,1}(j) + \alpha_{4,1}(j))}{|\alpha_{1,1}(1) + \alpha_{3,1}(1)|^2 + |\alpha_{2,1}(1) + \alpha_{4,1}(1)|^2}$$
$$B_{1m}(j) = \frac{(\alpha_{1,m}(j) + \alpha_{3,m}(j))(\alpha_{2,m}(1) + \alpha_{4,m}(1))^* - (\alpha_{2,m}(j) + \alpha_{4,m}(j))^*(\alpha_{1,m}(1) + \alpha_{3,m}(1))}{|\alpha_{1,m}(1) + \alpha_{3,m}(1)|^2 + |\alpha_{2,m}(1) + \alpha_{4,m}(1)|^2}$$
$$- \frac{(\alpha_{1,1}(j) + \alpha_{3,1}(j))(\alpha_{2,1}(1) + \alpha_{4,1}(1))^* - (\alpha_{2,1}(j) + \alpha_{4,1}(j))^*(\alpha_{1,1}(1) + \alpha_{3,1}(1))}{|\alpha_{1,1}(1) + \alpha_{3,1}(1)|^2 + |\alpha_{2,1}(1) + \alpha_{4,1}(1)|^2} \tag{2.39}$$

and

$$A_{2m}(j) = \frac{(\alpha_{1,m}(j) - \alpha_{3,m}(j))(\alpha_{1,m}(1) - \alpha_{3,m}(1))^* + (\alpha_{2,m}(j) - \alpha_{4,m}(j))^*(\alpha_{2,m}(j) - \alpha_{4,m}(j))}{|\alpha_{1,m}(1) - \alpha_{3,m}(1)|^2 + |\alpha_{2,m}(1) - \alpha_{4,m}(1)|^2}$$
$$- \frac{(\alpha_{1,1}(j) - \alpha_{3,1}(j))(\alpha_{1,1}(1) - \alpha_{3,1}(1))^* + (\alpha_{2,1}(j) - \alpha_{4,1}(j))^*(\alpha_{2,1}(j) - \alpha_{4,1}(j))}{|\alpha_{1,1}(1) - \alpha_{3,1}(1)|^2 + |\alpha_{2,1}(1) - \alpha_{4,1}(1)|^2}$$
$$B_{2m}(j) = \frac{(\alpha_{1,m}(j) - \alpha_{3,m}(j))(\alpha_{2,m}(1) - \alpha_{4,m}(1))^* - (\alpha_{2,m}(j) - \alpha_{4,m}(j))^*(\alpha_{1,m}(1) - \alpha_{3,m}(1))}{|\alpha_{1,m}(1) - \alpha_{3,m}(1)|^2 + |\alpha_{2,m}(1) - \alpha_{4,m}(1)|^2}$$
$$- \frac{(\alpha_{1,1}(j) - \alpha_{3,1}(j))(\alpha_{2,1}(1) - \alpha_{4,1}(1))^* - (\alpha_{2,1}(j) - \alpha_{4,1}(j))^*(\alpha_{1,1}(1) - \alpha_{3,1}(1))}{|\alpha_{1,1}(1) - \alpha_{3,1}(1)|^2 + |\alpha_{2,1}(1) - \alpha_{4,1}(1)|^2} \tag{2.40}$$

where m can be $2, \ldots, J + r - 1$.[4] The variance will be equal to $\frac{8}{SNR(|\alpha_{1,m}(1) \pm \alpha_{3,m}(1)|^2 + |\alpha_{2,m}(1) \pm \alpha_{4,m}(1)|^2)} + \frac{8}{SNR(|\alpha_{1,1}(1) \pm \alpha_{3,1}(1)|^2 + |\alpha_{2,1}(1) \pm \alpha_{4,1}(1)|^2)}$ respectively. Equations (2.37) and (2.38) have the same structure as (2.34) and (2.35). We note that $\mathcal{G}'$ is also multiple of a unitary matrix as[5] $\mathcal{G}$. It is as if, in (2.34) and (2.35) we have $J + r - 2$ receive antennas and $J - 1$ user groups. Therefore, we can separate and eliminate user No. 2 similar to user No. 1; except that the number of equivalent receive antennas will be one less, i.e. $(J + r - 2)$. Eliminating user No. 2 gives us $J + r - 3$ equations with $J - 2$ users. If we continue this procedure, we will have r equations for user No. J with no interference from other users. We can simply decode this user. Since we have r equations and our metric is ML, diversity gain will be equal to $4 \times r$. One should note that better performance is possible, if we use more complex decoding methods. For example, since we break down our code to two Alamouti structures, the MMSE-ZF method used in [15] for 2 transmit antenna systems can be applied here. This method that involves a matrix inversion, will provide a better error rate, compared with our simple ZF method. Also, one can decode users on the order of the strength of their respective channel and improve the performance. Our proposed algorithm can be combined with different decoding methods. The goal in here is to show it is possible to do multi-user detection for any number of transmit antennas,[6] using very few number of receive antennas. It is not meant to find the best performance, and there is no claim of optimality.

The flow of decoding is summarized below. In what follows we assume we want to decode user No. J; we can then simply change the ordering of users in case we are interested in decoding another user.

Algorithm

(1) Initialization. Let $i = 1$. Also, for all values of $j = 1, 2, \ldots, J$ and $m = 1, 2, \ldots, M$ let:

$$\begin{aligned} A_{1m}(j) &= \alpha_{1,m}(j) + \alpha_{3,m}(j) \\ B_{1m}(j) &= \alpha_{2,m}(j) + \alpha_{4,m}(j) \\ A_{2m}(j) &= \alpha_{1,m}(j) - \alpha_{3,m}(j) \quad \text{and} \\ B_{2m}(j) &= \alpha_{2,m}(j) - \alpha_{4,m}(j) \end{aligned}$$

Construct $\mathbf{R}'_{1,1}, \mathbf{R}'_{1,2}, \ldots, \mathbf{R}'_{1,M}$ and $\mathbf{R}'_{2,1}, \mathbf{R}'_{2,2}, \ldots, \mathbf{R}'_{2,M}$ from the M received signal vectors $\mathbf{R}_1, \mathbf{R}_2, \ldots, \mathbf{R}_M$.

(2) For all values of $j = i, i + 1, \ldots, J$ and $m = i, i + 1, \ldots, M$: if $i = 1$ let $\Omega_{1,m}(j) = \mathcal{G}(A_{1m}(j), B_{1m}(j))$ and $\Omega_{2,m}(j) = \mathcal{G}(A_{2m}(j), B_{2m}(j))$ else let $\Omega_{1,m}(j) = \mathcal{G}'(A_{1m}(j), B_{1m}(j))$ and $\Omega_{2,m}(j) = \mathcal{G}'(A_{2m}(j), B_{2m}(j))$.

[4] The distribution of noise will not be i.i.d. for $M > 2$. We describe this in the next chapter along with he optimal decoding method for it.

[5] One can easily show that had the channel matrix been $\mathcal{G}'$ instead of $\mathcal{G}$, we would have ended with $\mathcal{G}'$ as our equivalent channel, after suppressing the interference of the first user. Therefore, we will face $\mathcal{G}'$ at cancellation of all users except the first.

[6] As will be seen in the end of this section.

(3) For all values of $j = i+1, i+2, \ldots, J$ and $m = i+1, i+2, \ldots, M$:

$$\mathbf{R}'_{1,m} = \frac{\Omega^*_{1,m}(i)}{\frac{1}{2}\|\Omega_{1,m}(i)\|_F^2} \cdot \mathbf{R}'_{1,m} - \frac{\Omega^*_{1,i}(i)}{\frac{1}{2}\|\Omega_{1,i}(i)\|_F^2} \cdot \mathbf{R}'_{1,i}$$

$$\mathbf{R}'_{2,m} = \frac{\Omega^*_{2,m}(i)}{\frac{1}{2}\|\Omega_{2,m}(i)\|_F^2} \cdot \mathbf{R}'_{2,m} - \frac{\Omega^*_{2,i}(i)}{\frac{1}{2}\|\Omega_{2,i}(i)\|_F^2} \cdot \mathbf{R}'_{2,i}$$

if $i = 1$

$$A_{1m}(j) = \frac{A_{1m}(j)A^*_{1m}(i) + B^*_{1m}(j)B_{1m}(i)}{|A_{1m}(i)|^2 + |B_{1m}(i)|^2} - \frac{A_{1i}(j)A^*_{1i}(i) + B^*_{1i}(j)B_{1i}(i)}{|A_{1i}(i)|^2 + |B_{1i}(i)|^2}$$

$$B_{1m}(j) = \frac{A_{1m}(j)B^*_{1m}(i) - B^*_{1m}(j)A_{1m}(i)}{|A_{1m}(i)|^2 + |B_{1m}(i)|^2} - \frac{A_{1i}(j)B^*_{1i}(i) - B^*_{1i}(j)A_{1i}(i)}{|A_{1i}(i)|^2 + |B_{1i}(i)|^2}$$

$$A_{2m}(j) = \frac{A_{2m}(j)A^*_{2m}(i) + B^*_{2m}(j)B_{2m}(i)}{|A_{2m}(i)|^2 + |B_{2m}(i)|^2} - \frac{A_{2i}(j)A^*_{2i}(i) + B^*_{2i}(j)B_{2i}(i)}{|A_{2i}(i)|^2 + |B_{2i}(i)|^2}$$

$$B_{2m}(j) = \frac{A_{2m}(j)B^*_{2m}(i) - B^*_{2m}(j)A_{2m}(i)}{|A_{2m}(i)|^2 + |B_{2m}(i)|^2} - \frac{A_{2i}(j)B^*_{2i}(i) - B^*_{2i}(j)A_{2i}(i)}{|A_{2i}(i)|^2 + |B_{2i}(i)|^2}$$

else

$$A_{1m}(j) = \frac{A_{1m}(j)A^*_{1m}(i) + B_{1m}(j)B^*_{1m}(i)}{|A_{1m}(i)|^2 + |B_{1m}(i)|^2} - \frac{A_{1i}(j)A^*_{1i}(i) + B_{1i}(j)B^*_{1i}(i)}{|A_{1i}(i)|^2 + |B_{1i}(i)|^2}$$

$$B_{1m}(j) = \frac{-A_{1m}(j)B_{1m}(i) + B_{1m}(j)A_{1m}(i)}{|A_{1m}(i)|^2 + |B_{1m}(i)|^2} - \frac{-A_{1i}(j)B_{1i}(i) + B_{1i}(j)A_{1i}(i)}{|A_{1i}(i)|^2 + |B_{1i}(i)|^2}$$

$$A_{2m}(j) = \frac{A_{2m}(j)A^*_{2m}(i) + B_{2m}(j)B^*_{2m}(i)}{|A_{2m}(i)|^2 + |B_{2m}(i)|^2} - \frac{A_{2i}(j)A^*_{2i}(i) + B_{2i}(j)B^*_{2i}(i)}{|A_{2i}(i)|^2 + |B_{2i}(i)|^2}$$

$$B_{2m}(j) = \frac{-A_{2m}(j)B_{2m}(i) + B_{2m}(j)A_{2m}(i)}{|A_{2m}(i)|^2 + |B_{2m}(i)|^2} - \frac{-A_{2i}(j)B_{2i}(i) + B_{2i}(j)A_{2i}(i)}{|A_{2i}(i)|^2 + |B_{2i}(i)|^2}$$

(4) Let $i = i + 1$. If $i < J$ go to Step 2, otherwise stop the algorithm and decode the following system

$$\mathbf{R}'^T_{1,m} = (s_1(J) + s_3(J), s_2(J) + s_4(J)) \cdot \mathcal{G}'(A_{1m}(J), B_{1m}(J)) + \text{noise}$$

$$\mathbf{R}'^T_{2,m} = (s_1(J) - s_3(J), s_2(J) - s_4(J)) \cdot \mathcal{G}'(A_{2m}(J), B_{2m}(J)) + \text{noise}$$

for all $m = J, J+1, \ldots, M$. □

As can be noticed the complexity of this algorithm is linear in terms of the number of users and receive antennas. Also, the diversity provided in the last equation is $4 \times (M - J + 1)$ as discussed earlier.[7]

The relationship between the power of noise, at the beginning of each state and that of the end of the stage can be derived by comparing the ones in the first stage. Let us assume we are at Step i of detection and the noise power at the beginning of the step is σ^2. Based on what we found for the noise distribution in Step 2 and the recursive algorithm above, the noise power at the end of the step i at receive antenna m will be

$$\frac{\sigma^2}{|A_{1m}(i)|^2 + |B_{1m}(i)|^2} + \frac{\sigma^2}{|A_{1i}(i)|^2 + |B_{1i}(i)|^2} \tag{2.41}$$

One can use a similar approach and perform the task of interference cancellation for users equipped with 3 transmit antennas. We should simply remove one of the columns of the code matrix, e.g. fourth column, in encoding. In decoding, i.e. cancellation, we can use the above algorithm except that we put zero for every $\alpha_{4,m}(j)$. Note that the resulting code from column removal may not be optimal in terms of delay.

Theorem *Using modulation schemes in the form of* $\begin{pmatrix} A & B \\ B & A \end{pmatrix}$, *one can extend the above interference cancellation technique to users with any number of transmit antennas in the form of power of* 2.

Proof We use induction. Due to the structure of the code, we can convert the original 2^n transmit antenna system to two 2^{n-1} transmit antenna systems. Then, we use interference cancellation for each of those systems using $M = J$ receive antennas, and finally decode the two systems together. □

Corollary *We can extend this algorithm to users with any number of transmit antennas, by finding the first power of* 2 *not smaller than* N, *and the modulation scheme designed for that many transmit antennas using the method in the above theorem. Then, we can use the column removal method mentioned earlier to get the desired number of transmit antennas.*

4 Interference Cancellation Using Minimum Decoding Complexity Quasi-Orthogonal Space-Time Block Codes (MDC-QOSTBC)

In [28], a new Quasi-Orthogonal design has been introduced that trades a small amount of performance loss with simpler decoding. These codes, provide separate

[7]The diversity claims will be proved in the next chapter rigorously.

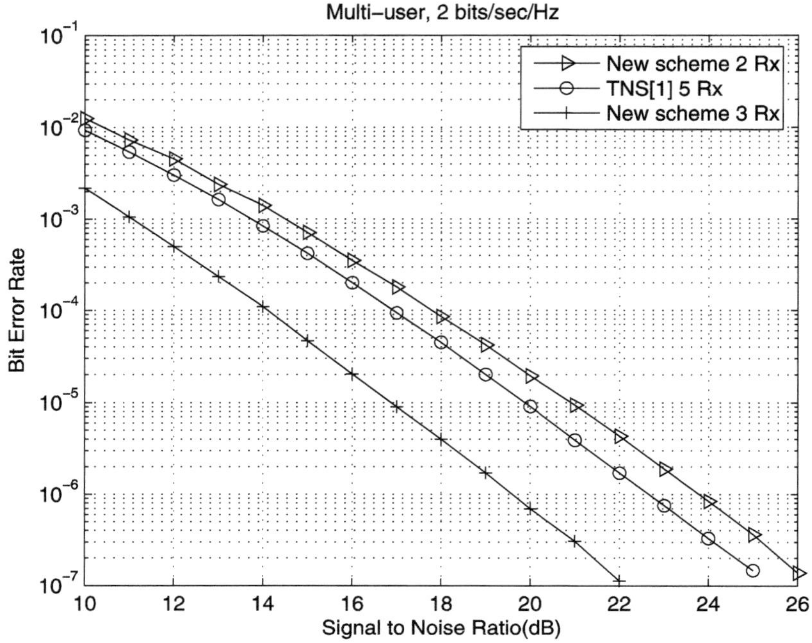

Fig. 2.1 Simulation results after interference cancellation when there are 2 users each transmitting QOSTBC with QPSK modulation

decoding complexity rather than pairwise, while maintaining full diversity and unit rate. In this section we briefly show that our introduced algorithm will work for this family of codes as well. The main idea is that, these codes can be presented as a mapping of the original QOSTBCs. We consider the code presented in [28] as an example

$$\mathbf{C}'(x_1,x_2,x_3,x_4) = \begin{pmatrix} x_1^R + jx_3^R & x_2^R + jx_4^R & -x_1^I + jx_3^I & -x_2^I + jx_4^I \\ -x_2^R + jx_4^R & x_1^R - jx_3^R & x_2^I + jx_4^I & -x_1^I - jx_3^I \\ -x_1^I + jx_3^I & -x_2^I + jx_4^I & x_1^R + jx_3^R & x_2^R + jx_4^r \\ x_2^I + jx_4^I & -x_1^I - jx_3^I & -x_2^R + jx_4^R & x_1^R - jx_3^R \end{pmatrix} \tag{2.42}$$

where $x_i^R = \Re\{x_i\}$ and $x_i^I = \Im\{x_i\}$. Recalling the definition of $\mathbf{C}$ used in the algorithm from (22), we realize that

$$\begin{aligned} \mathbf{C}'(x_1,x_2,x_3,x_4) &= \mathbf{C}(s_1,s_2,s_3,s_4) \\ s_1 &= x_1^R + jx_3^R \\ s_2 &= x_2^R + jx_4^R \end{aligned} \tag{2.43}$$

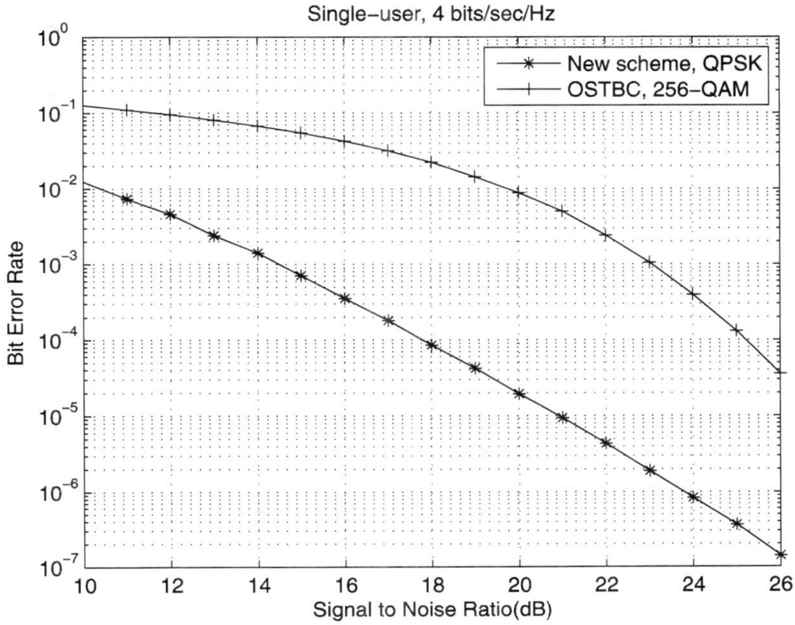

Fig. 2.2 Bit error probability vs. SNR for the new array processing scheme, and OSTBC at 2 bits/s/Hz; 8 transmit and 2 receive antennas

$$s_3 = -x_1^I + jx_3^I$$

$$s_4 = -x_2^I + jx_4^I$$

Therefore, in order to cancel out the interference of users equipped with MDC-QOSTBC, we will only need to map it to the corresponding QOSTBC. From there, we can apply the algorithm of previous section and detect the desired user. Then, we do the reverse mapping on the detected symbols and get the original ones.

5 Application of the New Interference Cancellation Scheme in Array Processing

Assume we have an interference cancellation scheme. The scheme does not put any assumption on the location of the users. Therefore, even if the users are all together, unified as part of a bigger user, the proposed decoding works and provides transmit diversity. If we unify J of such users, the spatial multiplexing of the overall system will be J times that of the individual users alone. In other words, if we have a system with JN transmit antennas, we can use a combination of J space-time codes with interference cancellation decoding, that sends J times more symbols per time slot

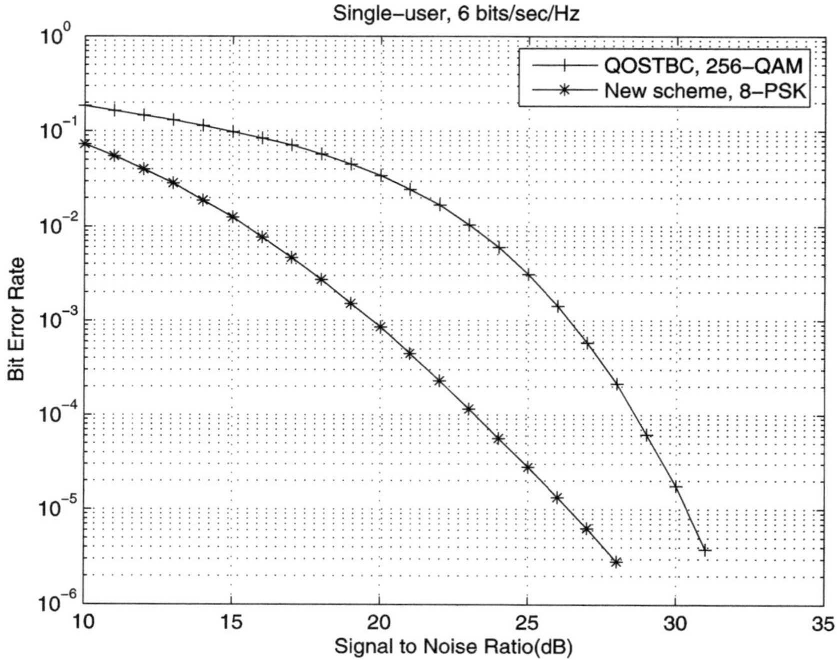

Fig. 2.3 Bit error probability vs. SNR for the new array processing scheme, and QOSTBC at 6 bits/s/Hz; 8 transmit and 2 receive antennas

than each of the individual space-time codes. This gives us an idea to apply the newly introduced scheme for increasing the rate in multiple antenna systems.

For example, let us assume we have 8 transmit antennas. We can always use a space-time code designed for 8 transmit antennas with full diversity, $8M$, and send at most one symbol per time slot [9]. Instead, we can exchange diversity for rate and use 2 separate QOSTBCs as mentioned in the previous section. This way we can send twice as many symbols and enjoy an acceptable diversity gain equal to $4(M-1)$.

As it is seen in the example above, our scheme provides a trade-off between rate and diversity. Since this scheme requires relatively less number of receive antennas, one may think of comparing its performance with popular multiple antenna schemes like OSTBC, QOSTBC, or BLAST. This task is performed in the next section.

6 Simulation Results

In this section we provide simulation results that confirm our analysis explained in the previous sections. The performance of our multi-user detection scheme is shown in Fig. 2.1. We consider 2 users each equipped with 4 antennas and transmitting

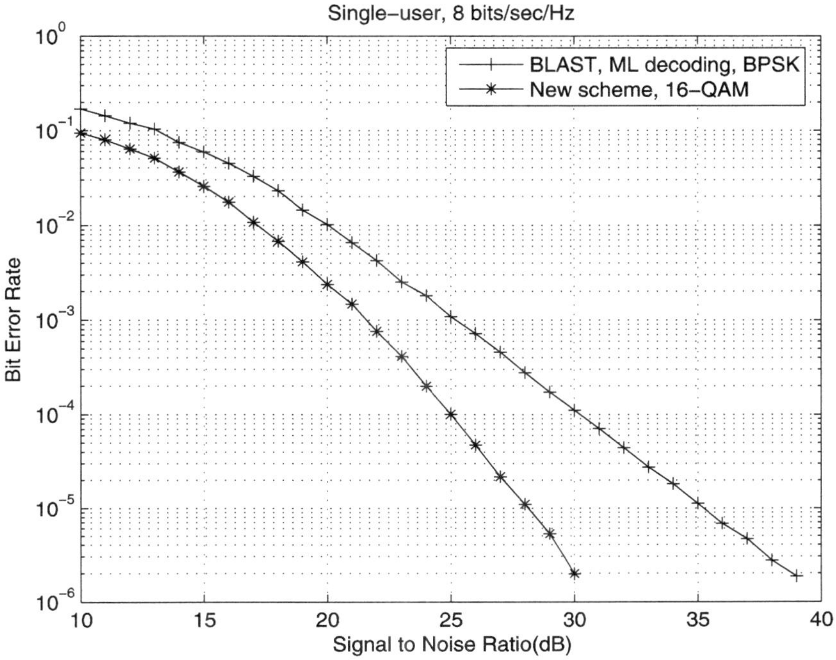

Fig. 2.4 Bit error probability vs. SNR for the new array processing scheme, and BLAST-ML at 8 bits/s/Hz; 8 transmit and 2 receive antennas

QOSTBCs. We compare this result with the one offered by [1] when it uses the same code, i.e. QOSTBC.[8] The constellation used is QPSK which provides a rate equal to 2 bits per channel use. The method offered in [1] requires at least 8 receive antennas for cancelling the interference whereas ours requires 2. It can be seen that our algorithm with usage of 3 receive antennas outperforms the old method when it uses 8 receive antennas.

Figures 2.2, 2.3, and 2.4 represent the comparison of the array processing scheme discussed in Sect. 5 with orthogonal space-time block codes, quasi-orthogonal space-time block codes, and BLAST respectively. We use ML for the BLAST code in Fig. 2.4. In Figs. 2.5 and 2.6, however, we are comparing the performance of our array processing scheme with ZF-BLAST with 8 and 12 antennas respectively. The array processing system in Figs. 2.1–2.5 consists of two 4-transmit antenna QOSTBC transmitters unified in one unit. In Fig. 2.2, we compare this system with the 16×8 orthogonal design in [11] and [25] using 256-QAM modulation. We use QPSK for the array processing scheme so that both systems provide a rate equal to 4 bits per channel use. In Fig. 2.3 we have used the 8-transmit antenna quasi-orthogonal design mentioned in [17] and [18], using rotation to provide full

[8]When [1] was published, QOSTBCs were not known; however, the method in that work allows usage of any modulation scheme including QOSTBCs.

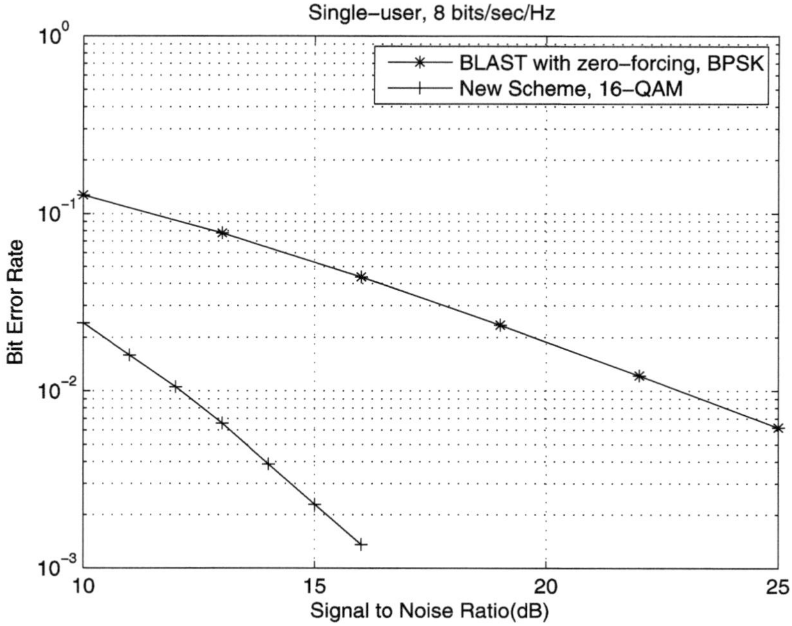

Fig. 2.5 Bit error probability vs. SNR for the new array processing scheme, and BLAST-ZF at 8 bits/s/Hz; 8 transmit and 8 receive antennas

diversity. For this experiment we have used 256-QAM for the 8-transmit antenna QOSTBC and 8PSK for the array processing system to provide the desired rate of 6 bits per channel use. In the case of BLAST in Fig. 2.4 we have used the regular 8-transmit antenna V-BLAST encoding with ML decoding. Note that the performance of the BLAST using the usual nulling and cancellation methods is much worse than that of BLAST using ML decoding. The constellation used for BLAST is BPSK and the one used for array processing system is 16-QAM; both systems transmit at a rate equal to 8 bits per channel use. In Fig. 2.5 both systems have 8 transmit and 8 receive antennas both using ZF as decoding method. The array processing system is using 16-QAM, and the BLAST code uses BPSK. Figure 2.6 shows comparison of a BLAST code with 12 transmit and receive antennas and the array processing system with same number of antennas. The array processing system consists of 3 unified 4 antenna users. Decoding method and constellation size is chosen the same as those of Fig. 2.5. For the experiments shown in Figs. 2.2, 2.3, and 2.4 we have considered 2 receive antennas at the receiver. In Figs. 2.5 and 2.6, however, we used the same number of receive antennas as that of the transmit antennas. The reason for this is that, many codes in the literature need at least as many receive antennas as the number of transmit antennas. This simulation gives a chance to compare the new scheme with those codes.

Note that both OSTBC and QOSTBC schemes have higher diversity order than our proposed scheme, and will take over at high enough Signal-to-Noise-Ratio.

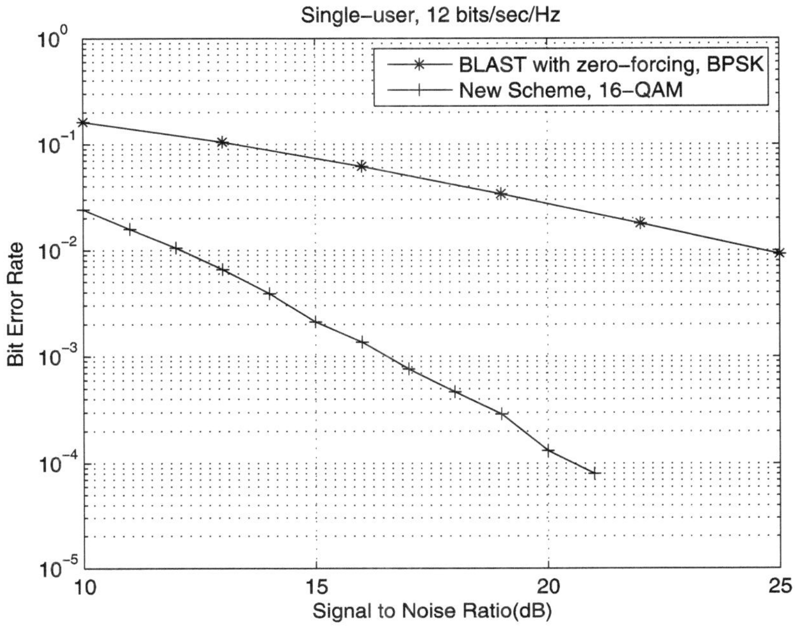

Fig. 2.6 Bit error probability vs. SNR for the new array processing scheme, and BLAST-ZF at 12 bits/s/Hz; 12 transmit and 12 receive antennas

However, our scheme performs better in a practical range of SNR and error rate. In the following, we summarize the main results of this chapter.

1. When there is $J > 2$ users each with $N > 2$ transmit antennas using a complex constellation, and orthogonal design, we show that one cannot perform interference cancellation with $M = J$ or less receive antennas.
2. When using a *real* constellation, we offer an interference cancellation technique that works for any number of users, J, having any number of transmit antennas; our algorithm only requires $M = J$ receive antennas, using real orthogonal designs.
3. For $N > 2$ transmit antenna equipped users using a complex constellation, we offer an interference cancellation technique that works for any number of users; again, our technique only requires receive antennas as many as the number of users.
4. We offer a joint array processing and space-time coding scheme for point to point communication (single user). The resulting code outperforms a number of popular modulation schemes, e.g. BLAST, OSBTC, and QOSTBC.

Chapter 3
Diversity Analysis of Multiple-Antenna Multi-User Systems

In this chapter we analyze the performance of the scheme we introduced in the previous chapter. We show that our intuitive guess about the diversity order of this method was correct. In other words, when J users each with N transmit antennas are sending data to a receiver with M antennas, the diversity order with usage of array processing will be $N(M - J + 1)$.

1 Diversity Order in a Communication Scheme

Diversity is usually defined as the exponent of the Signal-to-Noise-Ratio (SNR) in the error rate expression, high-SNR scenario [9],

$$d = -\lim_{SNR \to \infty} \frac{\log P_e}{\log SNR} \tag{3.1}$$

where P_e represents the probability of decoding error. One can derive other variants of diversity definition from the above formula. We mention one that will be used frequently in this chapter. In [12] the authors show that in every open-loop MIMO system, the error event is dominated by *Outage*. Outage is the scenario when the instantaneous SNR, due to bad channel realization, is unable to support the desired rate. The result from [12] states that

$$\lim_{SNR \to \infty} \frac{\log P_e}{\log SNR} = \lim_{SNR \to \infty} \frac{\log P_{\text{out}}}{\log SNR} \tag{3.2}$$

Therefore, when finding the diversity order, it is sufficient to know the outage behavior of the system [32]

$$d = \lim_{\epsilon \to 0^+} \frac{\log Pr\{\text{Instantaneous SNR} < \epsilon\}}{\log \epsilon} \tag{3.3}$$

S.J. Kazemitabar, *Coping with Interference in Wireless Networks*,
Signals and Communication Technology,
DOI 10.1007/978-90-481-9990-7_3,

2 Multi-User Detection Using Alamouti

Consider two users transmitting data simultaneously to a single receiver. Assume also, that they are using the Alamouti scheme. We denote the first user's message by $\mathbf{c} = (c_1, c_2)^T$, and the second user's message by $\mathbf{s} = (s_1, s_2)^T$. When using Alamouti the original code transmitted will be in the form of $\left(\begin{smallmatrix} c_1 & c_2 \\ -c_2^* & c_1^* \end{smallmatrix}\right)$ and $\left(\begin{smallmatrix} s_1 & s_2 \\ -s_2^* & s_1^* \end{smallmatrix}\right)$. As described in [29] however, one can derive an equivalent notation as following

$$\mathbf{r} = \mathbf{H} \cdot \mathbf{c} + \mathbf{G} \cdot \mathbf{s} + \mathbf{n} \tag{3.4}$$

where $\mathbf{r}$ has entries $\mathbf{r}_i = [\mathbf{r}_{1i} - \mathbf{r}_{2i}^*]^T$ with $\mathbf{r}_{1i}$ and $\mathbf{r}_{2i}$ being the signals received at the ith receive antenna over two consecutive symbol periods. $\mathbf{n}$ has a Gaussian distribution with $E[\mathbf{n}\mathbf{n}^*] = \frac{2}{SNR}\mathbf{I}$. Also, $\mathbf{H}$ and $\mathbf{G}$ are the equivalent channel matrices from the first and second user to the receiver respectively. Assuming 2 receive antennas, $\mathbf{H}$ and $\mathbf{G}$ will have an Alamouti structure as follows

$$\begin{aligned} &\mathbf{H} = \begin{pmatrix} \mathbf{H}_1 \\ \mathbf{H}_2 \end{pmatrix} \qquad \text{and} \quad \mathbf{G} = \begin{pmatrix} \mathbf{G}_1 \\ \mathbf{G}_2 \end{pmatrix} \\ &\mathbf{H}_i = \begin{pmatrix} h_{1i} & h_{2i} \\ -h_{2i}^* & h_{1i}^* \end{pmatrix} \quad \text{and} \quad \mathbf{G}_i = \begin{pmatrix} g_{1i} & g_{2i} \\ -g_{2i}^* & g_{1i}^* \end{pmatrix} \\ &i = 1, 2 \end{aligned} \tag{3.5}$$

In order to decode the message of each user one can use several techniques as mentioned in [15, 31]. The most trivial and computationally complex method is decoding both users together. This method, also known as ML, finds $\mathbf{c}$ and $\mathbf{s}$ as follows.

$$\operatorname{argmax} p(\mathbf{r}|\mathbf{c}, \mathbf{s}) = \frac{1}{\pi^2 \sigma^4} \exp\left(-\frac{1}{2\sigma^2} \|\mathbf{r} - \mathbf{Hc} - \mathbf{Gs}\|^2 \right) \tag{3.6}$$

The second method is Array-Processing (AP) and is sometimes named as Zero-Forcing (ZF) or soft interference cancellation. It requires very little computation and has linear decoding complexity. The following shows the process of this decoding method,

$$\begin{pmatrix} \mathbf{I}_2 & -\mathbf{G}_1\mathbf{G}_2^{-1} \\ -\mathbf{H}_2\mathbf{H}_1^{-1} & \mathbf{I}_2 \end{pmatrix} \begin{pmatrix} \mathbf{r}_1 \\ \mathbf{r}_2 \end{pmatrix} = \begin{pmatrix} \mathbf{H}' & 0 \\ 0 & \mathbf{G}' \end{pmatrix} \begin{pmatrix} \mathbf{c} \\ \mathbf{s} \end{pmatrix} + \begin{pmatrix} \mathbf{n}_1' \\ \mathbf{n}_2' \end{pmatrix} \tag{3.7}$$

Note that the inverse of the any Alamouti is a multiple of its Hermitian and therefore easy to compute.

In what follows, first, we prove a few lemmas that we use in the calculation of the diversity order in next section.

Lemma 1 *The following equality is valid for all* **H** *and* **G** *matrices of the form* (3.5):

$$
\begin{aligned}
&\|\mathbf{H}\|^2\|\mathbf{G}\|^2 - \|\mathbf{H}^{\dagger}\mathbf{G}\|^2 \\
&= \Bigg(a_5b_1 - a_6b_2 - a_7b_3 - a_8b_4 + a_1b_5 + a_2b_6 + a_3b_7 + a_4b_8 \\
&\quad - \frac{2(a_1b_4 + a_3b_2 + a_4b_1 - a_2b_3)(b_1b_8 + b_2b_7 - b_3b_6 + b_4b_5)}{b_1^2 + b_2^2 + b_3^2 + b_4^2} \Bigg)^2 \\
&\quad + \Bigg(a_6b_1 + a_5b_2 - a_8b_3 + a_7b_4 + a_1b_6 - a_2b_5 + a_3b_8 - a_4b_7 \\
&\quad - \frac{2(a_1b_4 + a_3b_2 + a_4b_1 - a_2b_3)(-b_1b_7 + b_2b_8 + b_3b_5 + b_4b_6)}{b_1^2 + b_2^2 + b_3^2 + b_4^2} \Bigg)^2 \\
&\quad + \Bigg(a_7b_1 + a_8b_2 + a_5b_3 - a_6b_4 + a_1b_7 - a_2b_8 - a_3b_5 + a_4b_6 \\
&\quad + \frac{2(a_1b_4 + a_3b_2 + a_4b_1 - a_2b_3)(-b_1b_6 + b_2b_5 - b_3b_8 - b_4b_7)}{b_1^2 + b_2^2 + b_3^2 + b_4^2} \Bigg)^2 \\
&\quad + \Bigg(a_8b_1 - a_7b_2 + a_6b_3 + a_5b_4 + a_1b_8 + a_2b_7 - a_3b_6 - a_4b_5 \\
&\quad + \frac{2(a_1b_4 + a_3b_2 + a_4b_1 - a_2b_3)(b_1b_5 + b_2b_6 + b_3b_7 - b_4b_8)}{b_1^2 + b_2^2 + b_3^2 + b_4^2} \Bigg)^2
\end{aligned} \tag{3.8}
$$

where,

$$
\begin{aligned}
h_{11} &= a_1 - ja_2, & h_{21} &= -a_3 + ja_4 \\
h_{12} &= -a_5 - ja_6, & h_{22} &= -a_7 - ja_8 \\
g_{11} &= b_1 + jb_2, & g_{21} &= b_3 + jb_4 \\
g_{12} &= b_5 + jb_6, & g_{22} &= b_7 - jb_8
\end{aligned} \tag{3.9}
$$

Proof See Appendix A. □

Lemma 2 *Assume a communication channel as* $Y = HX + N$ *with* $H = \alpha.\beta$, *where* α *and* β *are i.i.d. complex Gaussian random variables with zero mean and unit variance. The diversity order of this communication system is* 1.

Proof See Appendix A. □

Diversity Order of the ML Decoding

Consider the system described in (3.4) with M receive antennas. When using ML, the receiver finds the codeword that satisfies the minimum distance criterion for the following system

$$\begin{pmatrix} r_{11} & r_{12} & \cdots & r_{1M} \\ r_{21} & r_{22} & \cdots & r_{2M} \end{pmatrix} = \begin{pmatrix} c_1 & c_2 & s_1 & s_2 \\ -c_2^* & c_1^* & -s_2^* & s_1^* \end{pmatrix} \begin{pmatrix} h_{11} & h_{12} & \cdots & h_{1M} \\ h_{21} & h_{22} & \cdots & h_{2M} \\ g_{11} & g_{12} & \cdots & g_{1M} \\ g_{21} & g_{22} & \cdots & g_{2M} \end{pmatrix} + \begin{pmatrix} n_{11} & n_{12} & \cdots & n_{1M} \\ n_{21} & n_{22} & \cdots & n_{2M} \end{pmatrix} \tag{3.10}$$

The diversity of the above system is equal to the minimum rank of all the difference code matrices times the number of receive antennas [25]. For the above system this value will be $2M$. For more than two users, the diversity order will remain the same since the minimum rank does not change.[1]

Diversity Order of the Array-Processing Method

When there are two Alamouti-equipped transmitters, the effective SNR for user number one when using array-processing has been derived in [31] to be[2]

$$SNR_{\text{AP}} = \frac{\|\mathbf{H}\|^2}{\sigma^2}(1 - \|\Lambda\|^2) \tag{3.11}$$

where Λ is defined as

$$\Lambda = \frac{\mathbf{H}^\dagger \mathbf{G}}{\|\mathbf{H}\| \, \|\mathbf{G}\|} \tag{3.12}$$

[1]The rank of J concatenated Alamoutis-a $2J \times 2$ matrix-is always 2.

[2]In [31], the authors refer to this structure as Zero-Forcing (ZF).

$$
Pr\left\{\frac{\begin{array}{l}
(a_5b_1 - a_6b_2 - a_7b_3 - a_8b_4 + a_1b_5 + a_2b_6 + a_3b_7 + a_4b_8 \\
\quad - \frac{2(a_1b_4+a_3b_2+a_4b_1-a_2b_3)(b_1b_8+b_2b_7-b_3b_6+b_4b_5)}{b_1^2+b_3^2+b_4^2+b_2^2})^2 \\
\quad + (a_6b_1 + a_5b_2 - a_8b_3 + a_7b_4 + a_1b_6 - a_2b_5 + a_3b_8 - a_4b_7 \\
\quad - \frac{2(a_1b_4+a_3b_2+a_4b_1-a_2b_3)(-b_1b_7+b_2b_8+b_3b_5+b_4b_6)}{b_1^2+b_2^2+b_3^2+b_4^2})^2 \\
\quad + (a_7b_1 + a_8b_2 + a_5b_3 - a_6b_4 + a_1b_7 - a_2b_8 - a_3b_5 + a_4b_6 \\
\quad + \frac{2(a_1b_4+a_3b_2+a_4b_1-a_2b_3)(-b_1b_6+b_2b_5-b_3b_8-b_4b_7)}{b_1^2+b_2^2+b_3^2+b_4^2})^2 \\
\quad + (a_8b_1 - a_7b_2 + a_6b_3 + a_5b_4 + a_1b_8 + a_2b_7 - a_3b_6 - a_4b_5 \\
\quad + \frac{2(a_1b_4+a_3b_2+a_4b_1-a_2b_3)(b_1b_5+b_2b_6+b_3b_7-b_4b_8)}{b_1+b_2^2+b_3^2+b_4^2})^2
\end{array}}{\sigma^2(b_1^2 + b_2^2 + b_3^2 + b_4^2 + b_5^2 + b_6^2 + b_7^2 + b_8^2)} < \epsilon\right\}
$$

$$
= E_{\underline{\mathbf{b}}}\left[Pr\left\{\frac{\begin{array}{l}
(a_5b_1 - a_6b_2 - a_7b_3 - a_8b_4 + a_1b_5 + a_2b_6 + a_3b_7 + a_4b_8 \\
\quad - \frac{2(a_1b_4+a_3b_2+a_4b_1-a_2b_3)(b_1b_8+b_2b_7-b_3b_6+b_4b_5)}{b_1^2+b_3^2+b_4^2+b_2^2})^2 \\
\quad + (a_6b_1 + a_5b_2 - a_8b_3 + a_7b_4 + a_1b_6 - a_2b_5 + a_3b_8 - a_4b_7 \\
\quad - \frac{2(a_1b_4+a_3b_2+a_4b_1-a_2b_3)(-b_1b_7+b_2b_8+b_3b_5+b_4b_6)}{b_1^2+b_2^2+b_3^2+b_4^2})^2 \\
\quad + (a_7b_1 + a_8b_2 + a_5b_3 - a_6b_4 + a_1b_7 - a_2b_8 - a_3b_5 + a_4b_6 \\
\quad + \frac{2(a_1b_4+a_3b_2+a_4b_1-a_2b_3)(-b_1b_6+b_2b_5-b_3b_8-b_4b_7)}{b_1^2+b_2^2+b_3^2+b_4^2})^2 \\
\quad + (a_8b_1 - a_7b_2 + a_6b_3 + a_5b_4 + a_1b_8 + a_2b_7 - a_3b_6 - a_4b_5 \\
\quad + \frac{2(a_1b_4+a_3b_2+a_4b_1-a_2b_3)(b_1b_5+b_2b_6+b_3b_7-b_4b_8)}{b_1+b_2^2+b_3^2+b_4^2})^2
\end{array}}{b_1^2 + b_2^2 + b_3^2 + b_4^2 + b_5^2 + b_6^2 + b_7^2 + b_8^2} < \sigma^2\epsilon \;\middle|\; \underline{\mathbf{b}}\right\}\right] \tag{3.13}
$$

We now apply the formula in (3.3) to derive the diversity order.

$$d_{AP} = \lim_{\epsilon \to 0^+} \frac{\log Pr\{SNR_{AP} < \epsilon\}}{\log \epsilon} = \lim_{\epsilon \to 0^+} \frac{\log Pr\{\frac{\|\mathbf{H}\|^2.\|\mathbf{G}\|^2 - \|\mathbf{H}^\dagger \mathbf{G}\|^2}{\sigma^2 \|\mathbf{G}\|^2} < \epsilon\}}{\log \epsilon} \tag{3.14}$$

We can use (3.8) to simplify the numerator as shown in (3.13) on top of the next page, where $\underline{\mathbf{b}} = [b_1 b_2 \cdots b_8]$. In that equation, conditioned on $\underline{\mathbf{b}}$, each of the terms inside the 4 main parentheses is a zero-mean real Gaussian random variable due to independence of a_is. Once divided by the square root of the denominator their variance will become equal to one. Moreover, it can be easily checked that these Gaussian random variables are independent. Therefore, the sum of their squares is Chi-square distributed with 4 degrees of freedom and has the following density function

$$f(x) = xe^{-x} \quad x > 0 \tag{3.15}$$

For small enough ϵ,

$$\int_0^{\sigma^2 \epsilon} f(x)\,dx = \sigma^4 \epsilon^2 + O(\sigma^4 \epsilon^2) \tag{3.16}$$

where $f(x) = O(g(x))$ means there is a positive constant c such that $f(x) \leq cg(x)$ for the desired range of x. Since the quantity in (3.16) is independent of $\underline{\mathbf{b}}$, its expected value with respect to $\underline{\mathbf{b}}$ will remain the same. Therefore, we have

$$d = \lim_{\epsilon \to 0^+} \frac{\log(\sigma^4) + \log(\epsilon^2)}{\log(\epsilon)} = 2 \tag{3.17}$$

The Case with More than 2 Receive Antennas

Let us now assume the previous system with the exception that there are 3 receive antennas rather than two. For this system we have

$$\begin{aligned} \mathbf{r}_1 &= \mathbf{H}_1 \cdot \mathbf{c} + \mathbf{G}_1 \cdot \mathbf{s} + \mathbf{n}_1 \\ \mathbf{r}_2 &= \mathbf{H}_2 \cdot \mathbf{c} + \mathbf{G}_2 \cdot \mathbf{s} + \mathbf{n}_2 \\ \mathbf{r}_3 &= \mathbf{H}_3 \cdot \mathbf{c} + \mathbf{G}_3 \cdot \mathbf{s} + \mathbf{n}_3 \end{aligned} \tag{3.18}$$

After applying the array processing algorithm and cancelling the effect of user corresponding to message $\mathbf{s}$ we get

$$\begin{aligned} \mathbf{r}'_1 &= \left(\frac{\mathbf{G}_2^\dagger \mathbf{H}_2}{\|\mathbf{G}_2\|^2} - \frac{\mathbf{G}_1^\dagger \mathbf{H}_1}{\|\mathbf{G}_1\|^2} \right) \mathbf{c} + \mathbf{n}'_1 \\ \mathbf{r}'_2 &= \left(\frac{\mathbf{G}_3^\dagger \mathbf{H}_3}{\|\mathbf{G}_3\|^2} - \frac{\mathbf{G}_1^\dagger \mathbf{H}_1}{\|\mathbf{G}_1\|^2} \right) \mathbf{c} + \mathbf{n}'_2 \end{aligned} \tag{3.19}$$

Conditioned on $\mathbf{G}_i$s, the noise terms $\mathbf{n}'_1$ and $\mathbf{n}'_2$ are correlated Gaussian random variables. Similar statement applies to the new channel matrices $(\frac{\mathbf{G}_2^\dagger \mathbf{H}_2}{\|\mathbf{G}_2\|^2} - \frac{\mathbf{G}_1^\dagger \mathbf{H}_1}{\|\mathbf{G}_1\|^2})$ and $(\frac{\mathbf{G}_2^\dagger \mathbf{H}_2}{\|\mathbf{G}_2\|^2} - \frac{\mathbf{G}_1^\dagger \mathbf{H}_1}{\|\mathbf{G}_1\|^2})$. In [33] it is shown that in a Rayleigh fading system with receive correlation, like the one we have here, the diversity order will be NM as long as the correlation matrix of the channel is full-rank. Since, [33] assumes white noise, the equivalent correlation matrix in our case will be correlation matrix of the channel multiplied by the inverse of that of the noise. Clearly, the inverse of the correlation matrix of the noise accounts for the noise-whitening operation. Therefore, if we show that both of the these two correlation matrices are full-rank, we can conclude that the system in (3.19) provides a diversity order of 4 ($N = 2$, $M = 2$). The correlation matrix of noise is equal to

$$\begin{pmatrix} (\frac{\sigma^2}{\|\mathbf{G}_2\|^2} + \frac{\sigma^2}{\|\mathbf{G}_1\|^2})\mathbf{I}_2 & \frac{\sigma^2}{\|\mathbf{G}_1\|^2}\mathbf{I}_2 \\ \frac{\sigma^2}{\|\mathbf{G}_1\|^2}\mathbf{I}_2 & (\frac{\sigma^2}{\|\mathbf{G}_3\|^2} + \frac{\sigma^2}{\|\mathbf{G}_1\|^2})\mathbf{I}_2 \end{pmatrix} \tag{3.20}$$

where $\mathbf{I}_2$ is the 2×2 identity matrix. This matrix is clearly full-rank for almost (surely) all $\mathbf{G}_i$ realizations. It remains now to find the correlation matrix of the equivalent channel. Since both lines in (3.19) represent an Alamouti scheme, we can convert them back into the regular Alamouti representation as follows

$$\begin{aligned} \mathbf{y}_1 &= \begin{pmatrix} c_1 & c_2 \\ -c_2^* & c_1^* \end{pmatrix} \cdot \begin{pmatrix} A_1 + jA_2 \\ A_3 + jA_4 \end{pmatrix} + \text{noise} \\ \mathbf{y}_2 &= \begin{pmatrix} c_1 & c_2 \\ -c_2^* & c_1^* \end{pmatrix} \cdot \begin{pmatrix} B_1 + jB_2 \\ B_3 + jB_4 \end{pmatrix} + \text{noise} \end{aligned} \tag{3.21}$$

where the coefficients are normalized so that the noise terms have unit power. Using the SNR result from [31] and (3.8) we can write

$$A_1 = \frac{a_5b_1 - a_6b_2 - a_7b_3 - a_8b_4 + a_1b_5 + a_2b_6 + a_3b_7 + a_4b_8 - \frac{2(a_1b_4+a_3b_2+a_4b_1-a_2b_3)(b_1b_8+b_2b_7-b_3b_6+b_4b_5)}{b_1^2+b_2^2+b_3^2+b_4^2}}{\sigma\sqrt{b_1^2 + \cdots + b_8^2}}$$

$$A_2 = \frac{a_6b_1 + a_5b_2 - a_8b_3 + a_7b_4 + a_1b_6 - a_2b_5 + a_3b_8 - a_4b_7 - \frac{2(a_1b_4+a_3b_2+a_4b_1-a_2b_3)(-b_1b_7+b_2b_8+b_3b_5+b_4b_6)}{b_1^2+b_2^2+b_3^2+b_4^2}}{\sigma\sqrt{b_1^2 + \cdots + b_8^2}}$$

$$A_3 = \frac{a_7b_1 + a_8b_2 + a_5b_3 - a_6b_4 + a_1b_7 - a_2b_8 - a_3b_5 + a_4b_6 + \frac{2(a_1b_4+a_3b_2+a_4b_1-a_2b_3)(-b_1b_6+b_2b_5-b_3b_8-b_4b_7)}{b_1^2+b_2^2+b_3^2+b_4^2}}{\sigma\sqrt{b_1^2 + \cdots + b_8^2}}$$

$$A_4 = \frac{a_8b_1 - a_7b_2 + a_6b_3 + a_5b_4 + a_1b_8 + a_2b_7 - a_3b_6 - a_4b_5 + \frac{2(a_1b_4+a_3b_2+a_4b_1-a_2b_3)(b_1b_5+b_2b_6+b_3b_7-b_4b_8)}{b_1^2+b_2^2+b_3^2+b_4^2}}{\sigma\sqrt{b_1^2 + \cdots + b_8^2}}$$

$$B_1 = \frac{a_9b_1 - a_{10}b_2 - a_{11}b_3 - a_{12}b_4 + a_1b_9 + a_2b_{10} + a_3b_{11} + a_4b_{12} - \frac{2(a_1b_4+a_3b_2+a_4b_1-a_2b_3)(b_1b_{12}+b_2b_{11}-b_3b_{10}+b_4b_9)}{b_1^2+b_2^2+b_3^2+b_4^2}}{\sigma\sqrt{b_1^2 + \cdots + b_4^2 + b_9^2 + \cdots + b_{12}^2}}$$

$$B_2 = \frac{a_{10}b_1 + a_9b_2 - a_{12}b_3 + a_{11}b_4 + a_1b_{10} - a_2b_9 + a_3b_{12} - a_4b_{11} - \frac{2(a_1b_4+a_3b_2+a_4b_1-a_2b_3)(-b_1b_{11}+b_2b_{12}+b_3b_9+b_4b_{10})}{b_1^2+b_2^2+b_3^2+b_4^2}}{\sigma\sqrt{b_1^2+\cdots+b_4^2+b_9^2+\cdots+b_{12}^2}}$$

$$B_3 = \frac{a_{11}b_1 + a_{12}b_2 + a_9b_3 - a_{10}b_4 + a_1b_{11} - a_2b_{12} - a_3b_9 + a_4b_{10} + \frac{2(a_1b_4+a_3b_2+a_4b_1-a_2b_3)(-b_1b_{10}+b_2b_9-b_3b_{12}-b_4b_{11})}{b_1^2+b_2^2+b_3^2+b_4^2}}{\sigma\sqrt{b_1^2+\cdots+b_4^2+b_9^2+\cdots+b_{12}^2}}$$

$$B_4 = \frac{a_{12}b_1 - a_{11}b_2 + a_{10}b_3 + a_9b_4 + a_1b_{12} + a_2b_{11} - a_3b_{10} - a_4b_9 + \frac{2(a_1b_4+a_3b_2+a_4b_1-a_2b_3)(b_1b_9+b_2b_{10}+b_3b_{11}-b_4b_{12})}{b_1^2+b_2^2+b_3^2+b_4^2}}{\sigma\sqrt{b_1^2+\cdots+b_4^2+b_9^2+\cdots+b_{12}^2}} \tag{3.22}$$

The above values are real and imaginary parts of the channel coefficients. Instead of finding the complex correlation matrix we can find the following real correlation matrix

$$\mathbf{C} = E\{[\underline{\mathbf{A}}|\underline{\mathbf{B}}]^T[\underline{\mathbf{A}}|\underline{\mathbf{B}}]\} \tag{3.23}$$

where $\underline{\mathbf{A}} = [A_1\ A_2\ A_3\ A_4]$ and $\underline{\mathbf{B}} = [B_1\ B_2\ B_3\ B_4]$. It can be easily shown that if $\mathbf{C}$ is full-rank so will be the complex channel correlation matrix. We already know that $\{A_i\}$ and $\{B_i\}$ are each uncorrelated among themselves. Calculating $E\{A_iB_j\}$ we will have

$$\mathbf{C} = \frac{1}{\sigma^2}\begin{pmatrix} \mathbf{I} & \mathbf{X} \\ \mathbf{X} & \mathbf{I} \end{pmatrix} \tag{3.24}$$

where

$$X = \frac{1}{\sqrt{(b_1^2+\cdots+b_8^2)(b_1^2+\cdots+b_4^2+b_9^2+\cdots+b_{12}^2)}}\begin{pmatrix} b_5 & b_6 & b_7 & b_8 \\ b_6 & -b_5 & b_8 & -b_7 \\ b_7 & -b_8 & -b_5 & b_6 \\ b_8 & b_7 & -b_6 & -b_5 \end{pmatrix} \cdot \begin{pmatrix} b_9 & b_{10} & b_{11} & b_{12} \\ b_{10} & -b_9 & -b_{12} & b_{11} \\ b_{11} & b_{12} & -b_9 & -b_{10} \\ b_{12} & -b_{11} & b_{10} & -b_9 \end{pmatrix} \tag{3.25}$$

From [34] we have

$$\begin{aligned}\det(\mathbf{C}) &= \frac{1}{\sigma^{16}}\det(I - X^T \cdot X) \\ &= \frac{1}{\sigma^{16}}\left(1 - \frac{(b_5^2+\cdots+b_8^2)(b_9^2+\cdots+b_{12}^2)}{(b_1^2+\cdots+b_8^2)(b_1^2+\cdots+b_4^2+b_9^2+\cdots+b_{12}^2)}\right)\mathbf{I} \neq 0\end{aligned} \tag{3.26}$$

Therefore, the system described in (3.21) will provide full diversity, i.e. $2 \times 2 = 4$. This means that the described array processing scheme provides a diversity order equal to $N \times (M - J + 1)$ for the case of $N = 2$, $J = 2$, and $M = 3$.

We now further inspect the diversity order of the scheme by considering the general case of M receive antennas, while keeping the same number of users and transmit antennas. After cancelling the effect of user corresponding to message $\mathbf{s}$ we get

$$
\begin{aligned}
\mathbf{r}'_1 &= \left(\frac{\mathbf{G}_2^\dagger \mathbf{H}_2}{\|\mathbf{G}_2\|^2} - \frac{\mathbf{G}_1^\dagger \mathbf{H}_1}{\|\mathbf{G}_1\|^2}\right)\mathbf{c} + \mathbf{n}'_1 \\
\mathbf{r}'_2 &= \left(\frac{\mathbf{G}_3^\dagger \mathbf{H}_3}{\|\mathbf{G}_3\|^2} - \frac{\mathbf{G}_1^\dagger \mathbf{H}_1}{\|\mathbf{G}_1\|^2}\right)\mathbf{c} + \mathbf{n}'_2 \\
&\vdots \\
\mathbf{r}'_{M-1} &= \left(\frac{\mathbf{G}_M^\dagger \mathbf{H}_M}{\|\mathbf{G}_M\|^2} - \frac{\mathbf{G}_1^\dagger \mathbf{H}_1}{\|\mathbf{G}_1\|^2}\right)\mathbf{c} + \mathbf{n}'_{M-1}
\end{aligned}
\tag{3.27}
$$

We will again form the correlation matrix for noise and the equivalent Alamouti channel coefficients and examine whether they are full-rank. The noise correlation matrix will be

$$
\begin{pmatrix}
\left(\frac{\sigma^2}{\|\mathbf{G}_2\|^2} + \frac{\sigma^2}{\|\mathbf{G}_1\|^2}\right)\mathbf{I}_2 & \frac{\sigma^2}{\|\mathbf{G}_1\|^2}\mathbf{I}_2 & \cdots & \frac{\sigma^2}{\|\mathbf{G}_1\|^2}\mathbf{I}_2 \\
\frac{\sigma^2}{\|\mathbf{G}_1\|^2}\mathbf{I}_2 & \left(\frac{\sigma^2}{\|\mathbf{G}_3\|^2} + \frac{\sigma^2}{\|\mathbf{G}_1\|^2}\right)\mathbf{I}_2 & \cdots & \frac{\sigma^2}{\|\mathbf{G}_1\|^2}\mathbf{I}_2 \\
\vdots & \vdots & \ddots & \vdots \\
\frac{\sigma^2}{\|\mathbf{G}_1\|^2}\mathbf{I}_2 & \frac{\sigma^2}{\|\mathbf{G}_1\|^2}\mathbf{I}_2 & \cdots & \left(\frac{\sigma^2}{\|\mathbf{G}_2\|^2} + \frac{\sigma^2}{\|\mathbf{G}_M\|^2}\right)\mathbf{I}_2
\end{pmatrix}
\tag{3.28}
$$

which is equal to

$$
\begin{pmatrix}
\left(\frac{\sigma^2}{\|\mathbf{G}_2\|^2} + \frac{\sigma^2}{\|\mathbf{G}_1\|^2}\right) & \frac{\sigma^2}{\|\mathbf{G}_1\|^2} & \cdots & \frac{\sigma^2}{\|\mathbf{G}_1\|^2} \\
\frac{\sigma^2}{\|\mathbf{G}_1\|^2} & \left(\frac{\sigma^2}{\|\mathbf{G}_3\|^2} + \frac{\sigma^2}{\|\mathbf{G}_1\|^2}\right) & \cdots & \frac{\sigma^2}{\|\mathbf{G}_1\|^2} \\
\vdots & \vdots & \ddots & \vdots \\
\frac{\sigma^2}{\|\mathbf{G}_1\|^2} & \frac{\sigma^2}{\|\mathbf{G}_1\|^2} & \cdots & \left(\frac{\sigma^2}{\|\mathbf{G}_2\|^2} + \frac{\sigma^2}{\|\mathbf{G}_M\|^2}\right)
\end{pmatrix}
\otimes \mathbf{I}_2
\tag{3.29}
$$

The left hand side matrix is full-rank since it has $M-1$ nonzero eigenvalues as following[3]

$$
\frac{\sigma^2}{\|\mathbf{G}_2\|^2} + \frac{\sigma^2}{\|\mathbf{G}_1\|^2}, \quad \frac{\sigma^2}{\|\mathbf{G}_3\|^2} + \frac{\sigma^2}{\|\mathbf{G}_1\|^2}, \quad \ldots, \quad \frac{\sigma^2}{\|\mathbf{G}_M\|^2} + \frac{\sigma^2}{\|\mathbf{G}_1\|^2}
\tag{3.30}
$$

[3]The eigenvectors of this matrix are $\mathbf{e}_1, \mathbf{e}_2, \ldots, \mathbf{e}_{\mathbf{M-1}}$.

We should now examine the channel correlation matrix. In the general case of M receive antennas, we will have

$$\mathbf{C} = \frac{1}{\sigma_2}\begin{pmatrix} \mathbf{I} & \mathbf{X}_{12} & \cdots & \mathbf{X}_{1M-1} \\ \mathbf{X}_{21} & \mathbf{I} & \cdots & \mathbf{X}_{2M-1} \\ \vdots & \vdots & \ddots & \vdots \\ \mathbf{X}_{(M-1)1} & \mathbf{X}_{(M-1)2} & \cdots & \mathbf{X}_{(M-1)(M-1)} \end{pmatrix} \tag{3.31}$$

where $\mathbf{X}_{ij} = \mathbf{B}_i\mathbf{B}_j^T$ with

$$\mathbf{B}_i = \frac{1}{\sqrt{b_1^2 + \cdots + b_4^2 + b_{4i+1}^2 + \cdots + b_{4(i+1)}^2}} \cdot \begin{pmatrix} b_{4i+1} & b_{4i+2} & b_{4i+3} & b_{4(i+2)} \\ b_{4i+2} & -b_{4i+1} & b_{4(i+2)} & -b_{4i+3} \\ b_{4i+2} & -b_{4(i+2)} & -b_{4i+1} & b_{4i+2} \\ b_{4(i+1)} & b_{4i+3} & -b_{4i+2} & -b_{4i+1} \end{pmatrix} \tag{3.32}$$

It can be checked easily that $\mathbf{B}_i \cdot \mathbf{B}_i^T = \frac{b_{4i+1}^2+\cdots+b_{4(i+1)}^2}{b_1^2+\cdots+b_4^2+b_{4i+1}^2+\cdots+b_{4(i+1)}^2}\mathbf{I} = \beta_i\mathbf{I}$. It proves $\mathbf{C}$ full-rank if we can find $4(M-1)\times 4(M-1)$ matrix $\mathbf{U}$ such that

$$\mathbf{U}^T\mathbf{C}\mathbf{U} \tag{3.33}$$

has rank equal to $4(M-1)$. We will try to construct $\mathbf{U}$ based on the following structure

$$\mathbf{U} = \begin{pmatrix} \mathbf{u}_1 & |\cdots| & \mathbf{u}_{M-1} \end{pmatrix} \tag{3.34}$$

where $\mathbf{u}_i$s are $4(M-1)\times 4$. The following two lemmas will lead us construct the $\mathbf{U}$ matrix.

Lemma 3 *Given* $a_i = \frac{1}{\lambda^*+\beta_i-1}$ *where* λ^* *is a root of* $\sum_{i=1}^{M-1}\frac{\beta_i}{\lambda+\beta_i-1} = 1$ *we have*

$$C \cdot \begin{pmatrix} a_1\mathbf{B}_1 \\ a_2\mathbf{B}_2 \\ \vdots \\ a_{M-1}\mathbf{B}_{M-1} \end{pmatrix} = \lambda \begin{pmatrix} a_1\mathbf{B}_1 \\ a_2\mathbf{B}_2 \\ \vdots \\ a_{M-1}\mathbf{B}_{M-1} \end{pmatrix} \tag{3.35}$$

Proof See Appendix A. □

Lemma 4 *The following equation has* $M-1$ *distinct non-zero roots*

$$\sum_{i=1}^{M-1}\frac{\beta_i}{\lambda+\beta_i-1} = 1 \tag{3.36}$$

Proof See Appendix A. □

We name these distinct non-zero roots $\lambda_1^*, \ldots, \lambda_{M-1}^*$. Let us now define $\mathbf{u}_i$ vectors as following

$$\mathbf{u}_i = \begin{pmatrix} a_{1i}\mathbf{B}_1 \\ a_{2i}\mathbf{B}_2 \\ \vdots \\ a_{(M-1)i}\mathbf{B}_(M-1) \end{pmatrix} \tag{3.37}$$

where $a_{mi} = \frac{1}{\lambda_i^* + b_{m-1}}$ for $i, m = 1, \ldots, M-1$. From Lemma 3 and properties of $\mathbf{B}_j$s it is clear that

$$\begin{aligned} \mathbf{C} \cdot \mathbf{u}_i &= \lambda_i^* \mathbf{u}_i \\ \mathbf{u}_i^T \mathbf{u}_i &= \sum_j \beta_j a_{ji}^2 \mathbf{I} = S_i \mathbf{I} \end{aligned} \tag{3.38}$$

Also, since λ_i^*s are distinct we have

$$\begin{aligned} &\mathbf{u}_i^T \mathbf{C}\mathbf{u}_j = \mathbf{u}_i^T \lambda_j^* \mathbf{u}_j = \lambda_j^* \mathbf{u}_i^T \mathbf{u}_j \quad \text{and} \\ &\mathbf{u}_i^T \mathbf{C}\mathbf{u}_j = \mathbf{u}_i^T \mathbf{C}^T \mathbf{u}_j = (\mathbf{C}\mathbf{u}_i)^T \mathbf{u}_j = \lambda_i^* \mathbf{u}_i^T \mathbf{u}_j \quad \Longrightarrow \quad \mathbf{u}_i^T \mathbf{u}_j = 0 \quad \text{given } i \neq j. \end{aligned} \tag{3.39}$$

We are now ready to show why $\mathbf{C}$ is full-rank as follows

$$\begin{aligned} &\begin{pmatrix} \mathbf{u}_1^T \\ \mathbf{u}_2^T \\ \vdots \\ \mathbf{u}_{M-1}^T \end{pmatrix} \cdot \mathbf{C} \cdot \left(\mathbf{u}_1 \;\; | \;\; \mathbf{u}_2 \;\; | \;\; \cdots \;\; | \;\; \mathbf{u}_{M-1} \right) \\ &= \begin{pmatrix} \mathbf{u}_1^T \\ \mathbf{u}_2^T \\ \vdots \\ \mathbf{u}_{M-1}^T \end{pmatrix} \cdot \left(\mathbf{C}\mathbf{u}_1 \;\; | \;\; \mathbf{C}\mathbf{u}_2 \;\; | \;\; \cdots \;\; | \;\; \mathbf{C}\mathbf{u}_{M-1} \right) \\ &= \text{diag}\big(S_1\lambda_1^*, S_1\lambda_1^*, S_1\lambda_1^*, S_1\lambda_1^*, \ldots, S_{M-1}\lambda_M^*, S_{M-1}\lambda_{M-1}^*, S_{M-1}\lambda_{M-1}^*, \\ &\qquad S_{M-1}\lambda_{M-1}^*\big) \end{aligned} \tag{3.40}$$

which is clearly full-rank and it proves the same property for the matrix $\mathbf{C}$. Therefore, the channel correlation matrix is full-rank and the provided diversity for the scheme described in (3.27) is $2(M-1)$.

The Case with More than 2 Users

Let us now assume the multi-user system with 3 users and 3 antennas at the receiver as follows

$$\begin{aligned}\mathbf{r}_1 &= \mathbf{H}_1\cdot\mathbf{c}+\mathbf{G}_1\cdot\mathbf{s}+\mathbf{K}_1\cdot\mathbf{x}+\mathbf{n}_1\\ \mathbf{r}_2 &= \mathbf{H}_2\cdot\mathbf{c}+\mathbf{G}_2\cdot\mathbf{s}+\mathbf{K}_2\cdot\mathbf{x}+\mathbf{n}_2\\ \mathbf{r}_3 &= \mathbf{H}_3\cdot\mathbf{c}+\mathbf{G}_3\cdot\mathbf{s}+\mathbf{K}_3\cdot\mathbf{x}+\mathbf{n}_3\end{aligned} \quad (3.41)$$

Once we apply the cancellation technique on the user corresponding to message $\mathbf{x}$ we get

$$\begin{aligned}\mathbf{r}'_1 &= K_1^{-1}\mathbf{r}_1-\mathbf{K}_3^{-1}\mathbf{r}_3 = \left(\mathbf{K}_1^{-1}\mathbf{H}_1-\mathbf{K}_3^{-1}\mathbf{H}_3\right)\mathbf{c}+\left(\mathbf{K}_1^{-1}\mathbf{G}_1-\mathbf{K}_3^{-1}\mathbf{G}_3\right)\mathbf{s}+\mathbf{z}_1\\ \mathbf{r}'_2 &= K_2^{-1}\mathbf{r}_2-\mathbf{K}_3^{-1}\mathbf{r}_3 = \left(\mathbf{K}_2^{-1}\mathbf{H}_2-\mathbf{K}_3^{-1}\mathbf{H}_3\right)\mathbf{c}+\left(\mathbf{K}_2^{-1}\mathbf{G}_2-\mathbf{K}_3^{-1}\mathbf{G}_3\right)\mathbf{s}+\mathbf{z}_2\end{aligned} \quad (3.42)$$

We note that $\mathbf{K}_i^{-1} = \frac{\mathbf{K}_i^{\dagger}}{\|\mathbf{K}_i\|^2}$. Conditioned on $\mathbf{K}_i$s, the above system represents a Rayleigh fading channel with 2 users and 2 receive antennas. Therefore, similar to the system in (3.19) all the diversity claims of a 2 user systems (conditionally) apply.[4] In other words, the diversity order will be equal to 2. Taking the expectation over all $\mathbf{K}_i$s will not change this constant value and the diversity will remain 2. Similarly, when having M receive antennas for multi-user detection of 3 users we get diversity order of $2(M-3+1)$. Using induction on the number of users then, we can prove the following theorem.

Theorem 1 *Suppose we have J Alamouti-equipped users transmitting to the same receiver in the same frequency band and are time synchronized. Let us also assume that at the receiver we have M antennas and we use array processing as explained in* [30]. *The diversity provided to each user then will be equal to* $2(M-J+1)$.

3 Multi-User Detection for More than Two Transmit Antennas

In this section we first briefly explain the scheme in [30] and then find its provided diversity. Suppose, we have two users each with 4 transmit antennas equipped with QOSTBC. They are synchronously transmitting data to a receiver with two receive

[4] The only difference is that the noise and the channel coefficients are correlated. However, this will not affect the diversity results since the correlation matrices are exactly like those in (3.20) and (3.55) and therefore full-rank.

antennas as following

$$\begin{aligned}
\begin{pmatrix} r_{11} \\ r_{21} \\ r_{31} \\ r_{41} \end{pmatrix} &= \begin{pmatrix} c_1 & c_2 & c_3 & c_4 \\ -c_2^* & c_1^* & -c_4^* & c_3^* \\ c_3 & c_4 & c_1 & c_2 \\ -c_4^* & c_3^* & -c_2^* & c_1^* \end{pmatrix} \cdot \begin{pmatrix} h_{11} \\ h_{21} \\ h_{31} \\ h_{41} \end{pmatrix} \\
&\quad + \begin{pmatrix} s_1 & s_2 & s_3 & s_4 \\ -s_2^* & s_1^* & -s_4^* & s_3^* \\ s_3 & s_4 & s_1 & s_2 \\ -s_4^* & s_3^* & -s_2^* & s_1^* \end{pmatrix} \cdot \begin{pmatrix} g_{11} \\ g_{21} \\ g_{31} \\ g_{41} \end{pmatrix} + \begin{pmatrix} n_{11} \\ n_{21} \\ n_{31} \\ n_{41} \end{pmatrix} \\
\begin{pmatrix} r_{12} \\ r_{22} \\ r_{32} \\ r_{42} \end{pmatrix} &= \begin{pmatrix} c_1 & c_2 & c_3 & c_4 \\ -c_2^* & c_1^* & -c_4^* & c_3^* \\ c_3 & c_4 & c_1 & c_2 \\ -c_4^* & c_3^* & -c_2^* & c_1^* \end{pmatrix} \cdot \begin{pmatrix} h_{12} \\ h_{22} \\ h_{32} \\ h_{42} \end{pmatrix} \\
&\quad + \begin{pmatrix} s_1 & s_2 & s_3 & s_4 \\ -s_2^* & s_1^* & -s_4^* & s_3^* \\ s_3 & s_4 & s_1 & s_2 \\ -s_4^* & s_3^* & -s_2^* & s_1^* \end{pmatrix} \cdot \begin{pmatrix} g_{12} \\ g_{22} \\ g_{32} \\ g_{42} \end{pmatrix} + \begin{pmatrix} n_{12} \\ n_{22} \\ n_{32} \\ n_{42} \end{pmatrix}
\end{aligned} \tag{3.43}$$

We then define

$$\mathbf{r}_1 = \begin{pmatrix} r_{11} + r_{31} \\ -r_{21}^* - r_{41}^* \end{pmatrix}, \qquad \mathbf{r}_1' = \begin{pmatrix} r_{11} - r_{31} \\ -r_{21}^* + r_{41}^* \end{pmatrix} \tag{3.44}$$

Assuming similar definitions for $\mathbf{r}_2$ and $\mathbf{r}_2'$ we will have

$$\begin{aligned}
\mathbf{r}_1 &= \mathbf{H}_1\mathbf{c}^+ + \mathbf{G}_1\mathbf{s}^+ + \mathbf{n}_1, \qquad & \mathbf{r}_1' &= \mathbf{H}_1'\mathbf{c}^- + \mathbf{G}_1'\mathbf{s}^- + \mathbf{n}_1' \\
\mathbf{r}_2 &= \mathbf{H}_2\mathbf{c}^+ + \mathbf{G}_2\mathbf{s}^+ + \mathbf{n}_2, \qquad & \mathbf{r}_2' &= \mathbf{H}_1'\mathbf{c}^- + \mathbf{G}_1'\mathbf{s}^- + \mathbf{n}_2'
\end{aligned} \tag{3.45}$$

where

$$\begin{aligned}
\mathbf{H}_1 &= \begin{pmatrix} h_{11} + h_{31} & h_{21} + h_{41} \\ -h_{11}^* - h_{31}^* & h_{21}^* + h_{41}^* \end{pmatrix}, \qquad & \mathbf{H}_1' &= \begin{pmatrix} h_{11} - h_{31} & h_{21} - h_{41} \\ -h_{11}^* + h_{31}^* & h_{21}^* - h_{41}^* \end{pmatrix} \\
\mathbf{c}^+ &= \begin{pmatrix} c_1 + c_3 \\ c_2 + c_4 \end{pmatrix}, & \mathbf{c}^- &= \begin{pmatrix} c_1 - c_3 \\ c_2 - c_4 \end{pmatrix}
\end{aligned} \tag{3.46}$$

and the rest of the matrices are defined similarly.

Equation (3.45) reminds us of (3.4) and (3.5). Using the same array-processing algorithm one can cancel the effect of **s** and get the following

$$\begin{aligned}
\mathbf{r}^{+} &= \mathbf{G}_1^{-1}\mathbf{r}_1 - \mathbf{G}_2^{-1}\mathbf{r}_2 = \left(\mathbf{G}_1^{-1}\mathbf{H}_1 - \mathbf{G}_2^{-1}\mathbf{H}_2\right)\mathbf{c}^{+} + \left(\mathbf{G}_1^{-1}\mathbf{n}_1 - \mathbf{G}_2^{-1}\mathbf{n}_2\right) \\
&= \mathbf{A}\mathbf{s}^{+} + \mathbf{z} \\
\mathbf{r}^{-} &= \mathbf{G}_1'^{-1}\mathbf{r}_1' - \mathbf{G}_2'^{-1}\mathbf{r}_2' = \left(\mathbf{G}_1'^{-1}\mathbf{H}_1 - \mathbf{G}_2'^{-1}\mathbf{H}_2'\right)\mathbf{c}^{-} + \left(\mathbf{G}_1'^{-1}\mathbf{n}_1 - \mathbf{G}_2'^{-1}\mathbf{n}_2'\right) \\
&= \mathbf{A}'\mathbf{c}^{-} + \mathbf{z}'
\end{aligned} \tag{3.47}$$

where $\mathbf{A}$ and $\mathbf{A}'$ can be shown to be in the form of

$$\begin{aligned}
\mathbf{A} &= \begin{pmatrix} \alpha_1 & \alpha_2 \\ -\alpha_2^* & \alpha_1^* \end{pmatrix} \\
\mathbf{A}' &= \begin{pmatrix} \alpha_1' & \alpha_2' \\ -\alpha_2'^* & \alpha_1'^* \end{pmatrix}
\end{aligned} \tag{3.48}$$

Conditioned on $\mathbf{G}_i$ and $\mathbf{G}_i'$ values, the noise terms will be i.i.d. complex Gaussian random variables. Similar argument applies to α_i and α_i'. Now, if we perform the reverse of the conversion in (3.43)–(3.45) we get

$$\mathbf{R} = \begin{pmatrix} c_1 & c_2 & c_3 & c_4 \\ -c_2^* & c_1^* & -c_4^* & c_3^* \\ c_3 & c_4 & c_1 & c_2 \\ -c_4^* & c_3^* & -c_2^* & c_1^* \end{pmatrix} \begin{pmatrix} \frac{\alpha_1+\alpha_1'}{2} \\ \frac{\alpha_2+\alpha_2'}{2} \\ \frac{\alpha_1-\alpha_1'}{2} \\ \frac{\alpha_2-\alpha_2'}{2} \end{pmatrix} + \text{i.i.d. noise} \tag{3.49}$$

Conditioned on $\mathbf{G}_i$ and $\mathbf{G}_i'$ values, the above system is equivalent to a single-user QOSTBC with independent noise and Rayleigh fading channel coefficients. This system clearly provides a diversity order of four,[5] even after taking the expectation. Therefore, in the case of $J = 2$ users, $N = 4$ transmit, and $M = 2$ antennas the diversity order is $4 = N(M - J + 1)$.

The Case with More than 2 Receive Antennas

Let us consider the above system with the exception that there are three receive antennas instead of two. For this system we have

$$\begin{aligned}
\mathbf{r}_1 &= \mathbf{H}_1\mathbf{c}^{+} + \mathbf{G}_1\mathbf{s}^{+} + \mathbf{n}_1, & \mathbf{r}_1' &= \mathbf{H}_1'\mathbf{c}^{-} + \mathbf{G}_1'\mathbf{s}^{-} + \mathbf{n}_1' \\
\mathbf{r}_2 &= \mathbf{H}_2\mathbf{c}^{+} + \mathbf{G}_2\mathbf{s}^{+} + \mathbf{n}_2, & \mathbf{r}_2' &= \mathbf{H}_2'\mathbf{c}^{-} + \mathbf{G}_2'\mathbf{s}^{-} + \mathbf{n}_2' \\
\mathbf{r}_3 &= \mathbf{H}_3\mathbf{c}^{+} + \mathbf{G}_3\mathbf{s}^{+} + \mathbf{n}_3, & \mathbf{r}_3' &= \mathbf{H}_3'\mathbf{c}^{-} + \mathbf{G}_3'\mathbf{s}^{-} + \mathbf{n}_3'
\end{aligned} \tag{3.50}$$

[5] This is assuming rotated constellation for c_3 and c_4.

After canceling out $\mathbf{s}$ we get

$$
\begin{aligned}
\mathbf{r}_1^+ &= \mathbf{G}_2^{-1}\mathbf{r}_2 - \mathbf{G}_1^{-1}\mathbf{r}_1 = (\mathbf{G}_2^{-1}\mathbf{H}_2 - \mathbf{G}_1^{-1}\mathbf{H}_1)\mathbf{c}^+ + (\mathbf{G}_2^{-1}\mathbf{n}_2 - \mathbf{G}_1^{-1}\mathbf{n}_1)\\
&= \mathbf{A}_1\mathbf{s}^+ + \mathbf{z}_1\\
\mathbf{r}_1^- &= \mathbf{G}_1'^{-1}\mathbf{r}_1' - \mathbf{G}_2'^{-1}\mathbf{r}_2' = (\mathbf{G}_1'^{-1}\mathbf{H}_1 - \mathbf{G}_2'^{-1}\mathbf{H}_2')\mathbf{c}^- + (\mathbf{G}_1'^{-1}\mathbf{n}_1 - \mathbf{G}_2'^{-1}\mathbf{n}_2')\\
&= \mathbf{A}_1'\mathbf{c}^- + \mathbf{z}_1'\\
\mathbf{r}_2^+ &= \mathbf{G}_3^{-1}\mathbf{r}_3 - \mathbf{G}_1^{-1}\mathbf{r}_1 = (\mathbf{G}_3^{-1}\mathbf{H}_3 - \mathbf{G}_1^{-1}\mathbf{H}_1)\mathbf{c}^+ + (\mathbf{G}_3^{-1}\mathbf{n}_3 - \mathbf{G}_1^{-1}\mathbf{n}_1)\\
&= \mathbf{A}_2\mathbf{s}^+ + \mathbf{z}_2\\
\mathbf{r}_2^- &= \mathbf{G}_3'^{-1}\mathbf{r}_3' - \mathbf{G}_1'^{-1}\mathbf{r}_1' = (\mathbf{G}_3'^{-1}\mathbf{H}_3 - \mathbf{G}_1'^{-1}\mathbf{H}_1')\mathbf{c}^- + (\mathbf{G}_3'^{-1}\mathbf{n}_3 - \mathbf{G}_1'^{-1}\mathbf{n}_1')\\
&= \mathbf{A}_2'\mathbf{c}^- + \mathbf{z}_2'
\end{aligned}
\tag{3.51}
$$

where

$$
\begin{aligned}
\mathbf{A}_1 &= \begin{pmatrix} \alpha_{11} & \alpha_{21} \\ -\alpha_{21}^* & \alpha_{11}^* \end{pmatrix}, \qquad \mathbf{A}_1' = \begin{pmatrix} \alpha_{11}' & \alpha_{21}' \\ -\alpha_{21}'^* & \alpha_{11}'^* \end{pmatrix}\\
\mathbf{A}_2 &= \begin{pmatrix} \alpha_{12} & \alpha_{22} \\ -\alpha_{22}^* & \alpha_{12}^* \end{pmatrix}, \qquad \mathbf{A}_2' = \begin{pmatrix} \alpha_{12}' & \alpha_{22}' \\ -\alpha_{22}'^* & \alpha_{12}'^* \end{pmatrix}
\end{aligned}
\tag{3.52}
$$

Although Gaussian, neither the noise terms, nor the channel fades are uncorrelated. The correlation matrix for the $(\mathbf{z}_1\ \mathbf{z}_2)^T$ will be equal to

$$
\begin{pmatrix} (\frac{\sigma^2}{\|\mathbf{G}_2\|^2} + \frac{\sigma^2}{\|\mathbf{G}_1\|^2})\mathbf{I}_2 & \frac{\sigma^2}{\|\mathbf{G}_1\|^2}\mathbf{I}_2 \\ \frac{\sigma^2}{\|\mathbf{G}_1\|^2}\mathbf{I}_2 & (\frac{\sigma^2}{\|\mathbf{G}_3\|^2} + \frac{\sigma^2}{\|\mathbf{G}_1\|^2})\mathbf{I}_2 \end{pmatrix}
\tag{3.53}
$$

and for $(\mathbf{z}_1'\ \mathbf{z}_2')^T$ it will be

$$
\begin{pmatrix} (\frac{\sigma^2}{\|\mathbf{G}_2'\|^2} + \frac{\sigma^2}{\|\mathbf{G}_1'\|^2})\mathbf{I}_2 & \frac{\sigma^2}{\|\mathbf{G}_1'\|^2}\mathbf{I}_2 \\ \frac{\sigma^2}{\|\mathbf{G}_1'\|^2}\mathbf{I}_2 & (\frac{\sigma^2}{\|\mathbf{G}_3'\|^2} + \frac{\sigma^2}{\|\mathbf{G}_1'\|^2})\mathbf{I}_2 \end{pmatrix}
\tag{3.54}
$$

The correlation matrix of $(\mathbf{A}_1\ \mathbf{A}_2)$ and $(\mathbf{A}_1'\ \mathbf{A}_2')$ will be of the form

$$
\frac{1}{\sigma^2}\begin{pmatrix} \mathbf{I} & \mathbf{X} \\ \mathbf{X} & \mathbf{I} \end{pmatrix}
\tag{3.55}
$$

where $\mathbf{X}$ is in the form of[6]

$$\mathbf{X} = \frac{1}{\sqrt{(b_1^2+\cdots+b_8^2)(b_1^2+\cdots+b_4^2+b_9^2+\cdots+b_{12}^2)}} \begin{pmatrix} b_5 & b_6 & b_7 & b_8 \\ b_6 & -b_5 & b_8 & -b_7 \\ b_7 & -b_8 & -b_5 & b_6 \\ b_8 & b_7 & -b_6 & -b_5 \end{pmatrix}$$

$$\cdot \begin{pmatrix} b_9 & b_{10} & b_{11} & b_{12} \\ b_{10} & -b_9 & -b_{12} & b_{11} \\ b_{11} & b_{12} & -b_9 & -b_{10} \\ b_{12} & -b_{11} & b_{10} & -b_9 \end{pmatrix} \tag{3.56}$$

Clearly, all these correlation matrices are full-rank. Now, similar to the 2 receive antenna case, we can perform the reverse conversion and write the above equation in the following form

$$\mathbf{R} = \begin{pmatrix} c_1 & c_2 & c_3 & c_4 \\ -c_2^* & c_1^* & -c_4^* & c_3^* \\ c_3 & c_4 & c_1 & c_2 \\ -c_4^* & c_3^* & -c_2^* & c_1^* \end{pmatrix} \begin{pmatrix} \frac{\alpha_{11}+\alpha'_{11}}{2} & \frac{\alpha_{12}+\alpha'_{12}}{2} \\ \frac{\alpha_{21}+\alpha'_{21}}{2} & \frac{\alpha_{21}+\alpha'_{22}}{2} \\ \frac{\alpha_{11}-\alpha'_{11}}{2} & \frac{\alpha_{12}-\alpha'_{12}}{2} \\ \frac{\alpha_{21}-\alpha'_{21}}{2} & \frac{\alpha_{22}-\alpha'_{22}}{2} \end{pmatrix} + \text{noise} \tag{3.57}$$

The correlation matrix of the new channel and noise terms can be derived via row operations and block-concatenation of the correlation matrices in (3.53)–(3.56). Therefore, they will also be full-rank and the diversity order of the equivalent scheme shown in (3.57) will be $4 \times 2 = 8$.

For the general case of M receive antennas, one can perform similar operations and get to the following noise correlation

$$\begin{pmatrix} \left(\frac{\sigma^2}{\|\mathbf{G}_2\|^2}+\frac{\sigma^2}{\|\mathbf{G}_1\|^2}\right) & \frac{\sigma^2}{\|\mathbf{G}_1\|^2} & \cdots & \frac{\sigma^2}{\|\mathbf{G}_1\|^2} \\ \frac{\sigma^2}{\|\mathbf{G}_1\|^2} & \left(\frac{\sigma^2}{\|\mathbf{G}_3\|^2}+\frac{\sigma^2}{\|\mathbf{G}_1\|^2}\right) & \cdots & \frac{\sigma^2}{\|\mathbf{G}_1\|^2} \\ \vdots & \vdots & \ddots & \vdots \\ \frac{\sigma^2}{\|\mathbf{G}_1\|^2} & \frac{\sigma^2}{\|\mathbf{G}_1\|^2} & \cdots & \left(\frac{\sigma^2}{\|\mathbf{G}_2\|^2}+\frac{\sigma^2}{\|\mathbf{G}_M\|^2}\right) \end{pmatrix} \otimes \mathbf{I}_2 \tag{3.58}$$

and channel correlation

$$\mathbf{C} = \frac{1}{\sigma_2} \begin{pmatrix} \mathbf{I} & \mathbf{X}_{12} & \cdots & \mathbf{X}_{1M-1} \\ \mathbf{X}_{21} & \mathbf{I} & \cdots & \mathbf{X}_{2M-1} \\ \vdots & \vdots & \ddots & \vdots \\ \mathbf{X}_{(M-1)1} & \mathbf{X}_{(M-1)2} & \cdots & \mathbf{X}_{(M-1)M-1} \end{pmatrix} \tag{3.59}$$

---

[6] This just shows the structure. In order to rigorously calculate these terms refer to (3.21)–(3.25).

matrices. After reverse conversion, the equivalent single-user system will look like

$$\mathbf{R} = \begin{pmatrix} c_1 & c_2 & c_3 & c_4 \\ -c_2^* & c_1^* & -c_4^* & c_3^* \\ c_3 & c_4 & c_1 & c_2 \\ -c_4^* & c_3^* & -c_2^* & c_1^* \end{pmatrix} \begin{pmatrix} \frac{\alpha_{11}+\alpha'_{11}}{2} & \frac{\alpha_{12}+\alpha'_{12}}{2} & \cdots & \frac{\alpha_{1M-1}+\alpha'_{1M-1}}{2} \\ \frac{\alpha_{21}+\alpha'_{21}}{2} & \frac{\alpha_{22}+\alpha'_{22}}{2} & \cdots & \frac{\alpha_{2M-1}+\alpha'_{2M-1}}{2} \\ \frac{\alpha_{11}-\alpha'_{11}}{2} & \frac{\alpha_{12}-\alpha'_{12}}{2} & \cdots & \frac{\alpha_{1M-1}-\alpha'_{1M-1}}{2} \\ \frac{\alpha_{21}-\alpha'_{21}}{2} & \frac{\alpha_{22}-\alpha'_{22}}{2} & \cdots & \frac{\alpha_{2M-1}-\alpha'_{2M-1}}{2} \end{pmatrix} + \text{noise} \tag{3.60}$$

which provides the diversity order of $4(M-1)$ due to the full-rank correlation argument. Therefore, for the case of $J = 2$ users, $N = 4$ transmit antennas and general M receive antennas, the diversity order will be $N(M - J + 1)$.

The Case with More than 2 Users

Let us now assume the multi-user system with 3 users and 3 antennas at the receiver as follows

$$\begin{aligned}
\begin{pmatrix} r_{11} \\ r_{21} \\ r_{31} \\ r_{41} \end{pmatrix} &= \begin{pmatrix} c_1 & c_2 & c_3 & c_4 \\ -c_2^* & c_1^* & -c_4^* & c_3^* \\ c_3 & c_4 & c_1 & c_2 \\ -c_4^* & c_3^* & -c_2^* & c_1^* \end{pmatrix} \cdot \begin{pmatrix} h_{11} \\ h_{21} \\ h_{31} \\ h_{41} \end{pmatrix} + \begin{pmatrix} s_1 & s_2 & s_3 & s_4 \\ -s_2^* & s_1^* & -s_4^* & s_3^* \\ s_3 & s_4 & s_1 & s_2 \\ -s_4^* & s_3^* & -s_2^* & s_1^* \end{pmatrix} \cdot \begin{pmatrix} g_{11} \\ g_{21} \\ g_{31} \\ g_{41} \end{pmatrix} \\
&+ \begin{pmatrix} x_1 & x_2 & x_3 & x_4 \\ -x_2^* & x_1^* & -x_4^* & x_3^* \\ x_3 & s_4 & X_1 & s_2 \\ -x_4^* & x_3^* & -x_2^* & x_1^* \end{pmatrix} \cdot \begin{pmatrix} k_{11} \\ k_{21} \\ k_{31} \\ k_{41} \end{pmatrix} + \begin{pmatrix} n_{11} \\ n_{21} \\ n_{31} \\ n_{41} \end{pmatrix} \\
\begin{pmatrix} r_{12} \\ r_{22} \\ r_{32} \\ r_{42} \end{pmatrix} &= \begin{pmatrix} c_1 & c_2 & c_3 & c_4 \\ -c_2^* & c_1^* & -c_4^* & c_3^* \\ c_3 & c_4 & c_1 & c_2 \\ -c_4^* & c_3^* & -c_2^* & c_1^* \end{pmatrix} \cdot \begin{pmatrix} h_{12} \\ h_{22} \\ h_{32} \\ h_{42} \end{pmatrix} + \begin{pmatrix} s_1 & s_2 & s_3 & s_4 \\ -s_2^* & s_1^* & -s_4^* & s_3^* \\ s_3 & s_4 & s_1 & s_2 \\ -s_4^* & s_3^* & -s_2^* & s_1^* \end{pmatrix} \cdot \begin{pmatrix} g_{12} \\ g_{22} \\ g_{32} \\ g_{42} \end{pmatrix} \\
&+ \begin{pmatrix} x_1 & x_2 & x_3 & x_4 \\ -x_2^* & x_1^* & -x_4^* & x_3^* \\ x_3 & s_4 & X_1 & s_2 \\ -x_4^* & x_3^* & -x_2^* & x_1^* \end{pmatrix} \cdot \begin{pmatrix} k_{12} \\ k_{22} \\ k_{32} \\ k_{42} \end{pmatrix} + \begin{pmatrix} n_{12} \\ n_{22} \\ n_{32} \\ n_{42} \end{pmatrix} \\
\begin{pmatrix} r_{13} \\ r_{23} \\ r_{33} \\ r_{43} \end{pmatrix} &= \begin{pmatrix} c_1 & c_2 & c_3 & c_4 \\ -c_2^* & c_1^* & -c_4^* & c_3^* \\ c_3 & c_4 & c_1 & c_2 \\ -c_4^* & c_3^* & -c_2^* & c_1^* \end{pmatrix} \cdot \begin{pmatrix} h_{13} \\ h_{23} \\ h_{33} \\ h_{43} \end{pmatrix} + \begin{pmatrix} s_1 & s_2 & s_3 & s_4 \\ -s_2^* & s_1^* & -s_4^* & s_3^* \\ s_3 & s_4 & s_1 & s_2 \\ -s_4^* & s_3^* & -s_2^* & s_1^* \end{pmatrix} \cdot \begin{pmatrix} g_{13} \\ g_{23} \\ g_{33} \\ g_{43} \end{pmatrix} \\
&+ \begin{pmatrix} x_1 & x_2 & x_3 & x_4 \\ -x_2^* & x_1^* & -x_4^* & x_3^* \\ x_3 & s_4 & X_1 & s_2 \\ -x_4^* & x_3^* & -x_2^* & x_1^* \end{pmatrix} \cdot \begin{pmatrix} k_{13} \\ k_{23} \\ k_{33} \\ k_{43} \end{pmatrix} + \begin{pmatrix} n_{13} \\ n_{23} \\ n_{33} \\ n_{43} \end{pmatrix}
\end{aligned} \tag{3.61}$$

Now, if we perform the conversion to Alamouti form we get to the following

$$\begin{aligned}
\mathbf{r}_1 &= \mathbf{H}_1\mathbf{c}^+ + \mathbf{G}_1\mathbf{s}^+ + \mathbf{n}_1, & \mathbf{r}'_1 &= \mathbf{H}'_1\mathbf{c}^- + \mathbf{G}'_1\mathbf{s}^- + \mathbf{K}'_1\mathbf{x}^- + \mathbf{n}'_1 \\
\mathbf{r}_2 &= \mathbf{H}_2\mathbf{c}^+ + \mathbf{G}_2\mathbf{s}^+ + \mathbf{n}_2, & \mathbf{r}'_2 &= \mathbf{H}'_2\mathbf{c}^- + \mathbf{G}'_2\mathbf{s}^- + \mathbf{K}'_2\mathbf{x}^- + \mathbf{n}'_2 \\
\mathbf{r}_3 &= \mathbf{H}_3\mathbf{c}^+ + \mathbf{G}_3\mathbf{s}^+ + \mathbf{n}_3, & \mathbf{r}'_3 &= \mathbf{H}'_3\mathbf{c}^- + \mathbf{G}'_3\mathbf{s}^- + \mathbf{K}'_3\mathbf{x}^- + \mathbf{n}'_3
\end{aligned} \tag{3.62}$$

Once we apply the cancellation technique on the user corresponding to message $\mathbf{x}$ we get

$$\begin{aligned}
\mathbf{r}_1^+ &= \mathbf{K}_2^{-1}\mathbf{r}_2 - \mathbf{K}_1^{-1}\mathbf{r}_1 = \mathbf{A}_1\mathbf{s}^+ + \mathbf{B}_1\mathbf{x}^+ + \mathbf{z}_1 \\
\mathbf{r}_1^- &= \mathbf{K}_1'^{-1}\mathbf{r}'_1 - \mathbf{K}_2'^{-1}\mathbf{r}'_2 = \mathbf{A}'_1\mathbf{c}^- + \mathbf{B}'_1\mathbf{x}^- + \mathbf{z}'_1 \\
\mathbf{r}_2^+ &= \mathbf{K}_3^{-1}\mathbf{r}_3 - \mathbf{K}_1^{-1}\mathbf{r}_1 = \mathbf{A}_2\mathbf{s}^+ + \mathbf{B}_2\mathbf{x}^+ + \mathbf{z}_2 \\
\mathbf{r}_2^- &= \mathbf{K}_3'^{-1}\mathbf{r}'_3 - \mathbf{K}_1'^{-1}\mathbf{r}'_1 = \mathbf{A}'_2\mathbf{c}^- + \mathbf{B}'_2\mathbf{x}^- + \mathbf{z}'_2
\end{aligned} \tag{3.63}$$

Conditioned on $\mathbf{K}_i$s and $\mathbf{K}'_i$s, the above system represents a Rayleigh fading channel with 2 users and 2 receive antennas. Therefore, similar to the system in (3.19) all the diversity claims of a 2 user systems (conditionally) apply.[7] In other words, the diversity order will be equal to 4. Taking the expectation over all $\mathbf{K}_i$s and $\mathbf{K}'_i$s will not change this constant value and the diversity will remain 4. Similarly, when having M receive antennas for multi-user detection of 3 users we get diversity order of $4(M - 3 + 1)$. Using induction on the number of users then, we can prove the following theorem.

Theorem 2 *Suppose we have J QOSTBC-equipped users transmitting to the same receiver in the same frequency band and are time synchronized. Let us also assume that at the receiver we have M antennas and we use array processing as explained in the previous chapter. The diversity provided to each user then will be equal to* $4(M - J + 1)$.

In the previous chapter we showed how, using $\left(\begin{smallmatrix}\mathbf{A} & \mathbf{B}\\ \mathbf{B} & \mathbf{A}\end{smallmatrix}\right)$ one can generalize the array processing method to all number of transmit antennas. The trick when $N = 2^k$ is to break the system into two systems with $N = 2^{k-1}$ and then perform the interference cancellation technique on each of them separately. Then, one can combine them back and get to the original system. Similar to the method we showed for converting the $N = 4$ to $N = 2$ systems, one can perform the same diversity analysis on any $N = 2^k$ transmit antenna system with $\left(\begin{smallmatrix}\mathbf{A} & \mathbf{B}\\ \mathbf{B} & \mathbf{A}\end{smallmatrix}\right)$ structure. In addition, the result can be extended to non-power-of-2s with column removal method to form the following corollary.

[7] The only difference is that the noise and the channel coefficients are correlated. However, this will not affect the diversity results since the correlation matrices are exactly like those in (3.56) and (3.58) and therefore full-rank.

Corollary *Assume we have J users each with N transmit antennas using the $\binom{\mathbf{A}\ \mathbf{B}}{\mathbf{B}\ \mathbf{A}}$ structure explained above. They are all sending data synchronously to a receive with $M \geq J$ receive antennas. The diversity of the array processing method will be equal to $N(M - J + 1)$.*

4 Joint Array Processing and Space-Time Coding

Suppose we have an interference cancellation scheme. The scheme does not put any assumption on the location of the users. Therefore, even if the users are all together, unified as part of a bigger user, the proposed decoding works and provides transmit diversity. If we unify J of such users, the number of independent streams of the overall system will be J times that of the individual users alone. In other words, if we have a system with $2J$ transmit antennas, we can use a combination of J space-time codes with interference cancellation, that sends J times more symbols per time slot than each of the individual space-time codes. It was shown in the previous chapter that these type of codes outperform several classes of known space-time codes in a large range of SNRs.

The diversity order of these codes will be equal to the diversity order of each of sub-systems used within, due to symmetry of the problem. Therefore, if for example, we use two combined Alamouti codes in a 4-transmit antenna system, with 2 receive antennas and use array-processing at the receiver, the diversity will be 2.

Similarly, in ML method the diversity order J combined N transmit antenna systems will be NM where M is the number of total receive antennas.

5 Discussion

Our diversity analysis concurs with the simulation results provided in Fig. 3.1. Intuitively, one can think of one of the $J - 1$ of the receive antennas as the canceler and the rest as the combiner. The diversity of a 2 Tx by $(M - J + 1)$ Rx system then, is expected to be equal to $2(M - J + 1)$. In the scheme proposed by Tarokh et al. [1] which is similar to VBLAST, the diversity for different users (different streams) is unequal. First user has the highest diversity and the last user has the least. In the system that we described all users (streams) have the same diversity.

Appendix A

Proof of Lemma 1 The equality can be checked easily after applying the variable change. Here, we show the steps how to derive it.

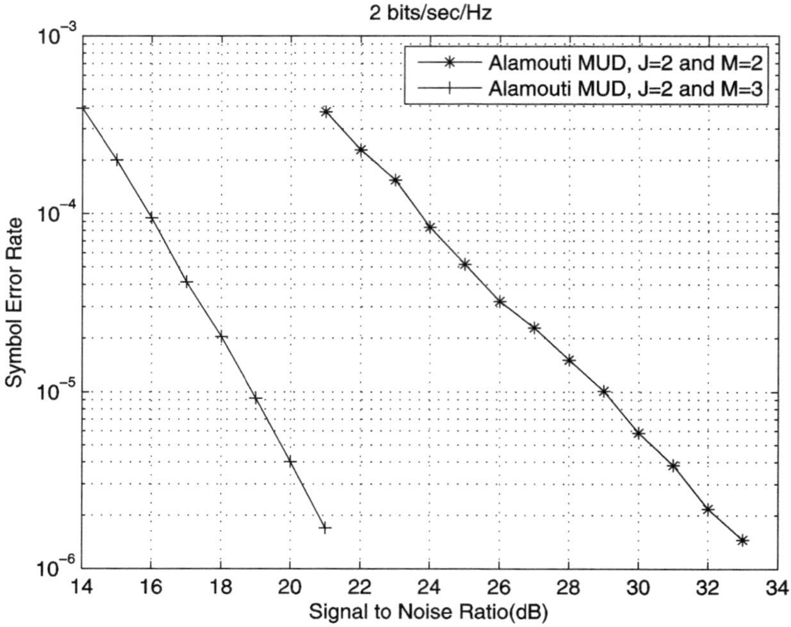

Fig. 3.1 Comparison of the two MUD systems with two Alamouti equipped users

The goal is to represent $\|\mathbf{H}\|^2\|\mathbf{G}\|^2 - \|\mathbf{H}^\dagger\mathbf{G}\|^2$ in terms of sum of square terms. For that purpose, let

$$\begin{aligned}
h_{11} &= a_1 - ja_2, & h_{21} &= -a_3 + ja_4 \\
h_{12} &= -a_5 - ja_6, & h_{22} &= -a_7 - ja_8 \\
g_{11} &= b_1 + jb_2, & g_{21} &= b_3 + jb_4 \\
g_{12} &= b_5 + jb_6, & g_{22} &= b_7 - jb_8
\end{aligned} \tag{3.64}$$

We use Degen's eight-square identity [35]

$$\begin{aligned}
&\left(a_1^2 + a_2^2 + a_3^2 + a_4^2 + a_5^2 + a_6^2 + a_7^2 + a_8^2\right)\left(b_1^2 + b_2^2 + b_3^2 + b_4 + b_5^2 + b_6^2 + b_7^2 + b_8^2\right) \\
&= (a_1b_1 - a_2b_2 - a_3b_3 - a_4b_4 - a_5b_5 - a_6b_6 - a_7b_7 - a_8b_8)^2 \\
&\quad + (a_2b_1 + a_1b_2 + a_4b_3 - a_3b_4 + a_6b_5 - a_5b_6 - a_8b_7 + a_7b_8)^2 \\
&\quad + (a_3b_1 - a_4b_2 + a_1b_3 + a_2b_4 + a_7b_5 + a_8b_6 - a_5b_7 - a_6b_8)^2 \\
&\quad + (a_4b_1 + a_3b_2 - a_2b_3 + a_1b_4 + a_8b_5 - a_7b_6 + a_6b_7 - a_5b_8)^2 \\
&\quad + (a_5b_1 - a_6b_2 - a_7b_3 - a_8b_4 + a_1b_5 + a_2b_6 + a_3b_7 + a_4b_8)^2 \\
&\quad + (a_6b_1 + a_5b_2 - a_8b_3 + a_7b_4 - a_2b_5 + a_1b_6 - a_4b_7 + a_3b_8)^2
\end{aligned}$$

$$+ (a_7b_1 + a_8b_2 + a_5b_3 - a_6b_4 - a_3b_5 + a_4b_6 + a_1b_7 - a_2b_8)^2$$
$$+ (a_8b_1 - a_7b_2 + a_6b_3 + a_5b_4 - a_4b_5 - a_3b_6 + a_2b_7 + a_1b_8)^2 \tag{3.65}$$

After simplification we get

$$\begin{aligned}
&\|\mathbf{H}\|^2\|\mathbf{G}\|^2 - \|\mathbf{H}^\dagger\mathbf{G}\|^2 \\
&= (a_5b_1 - a_6b_2 - a_7b_3 - a_8b_4 + a_1b_5 + a_2b_6 + a_3b_7 + a_4b_8)^2 \\
&\quad + (a_6b_1 + a_5b_2 - a_8b_3 + a_7b_4 - a_2b_5 + a_1b_6 - a_4b_7 + a_3b_8)^2 \\
&\quad + (a_7b_1 + a_8b_2 + a_5b_3 - a_6b_4 - a_3b_5 + a_4b_6 + a_1b_7 - a_2b_8)^2 \\
&\quad + (a_8b_1 - a_7b_2 + a_6b_3 + a_5b_4 - a_4b_5 - a_3b_6 + a_2b_7 + a_1b_8)^2 \\
&\quad + 4(a_1b_4 + a)3b_2 + a4b_1 - a_2b_3).(a_8b_5 - a_7b_6 + a_6b_7 - a_5b_8)
\end{aligned} \tag{3.66}$$

Let us define $A_1, A_2, A_3, A_4, B_1, B_2, B_3, B_4$ as follows,

$$\begin{aligned}
\begin{pmatrix} A_1 \\ A_2 \\ A_3 \\ A_4 \end{pmatrix} &= \begin{pmatrix} a_5 & -a_6 & -a_7 & -a_8 \\ a_6 & a_5 & -a_8 & a_7 \\ a_7 & a_8 & a_5 & -a_6 \\ a_8 & -a_7 & a_6 & a_5 \end{pmatrix} \cdot \begin{pmatrix} b_1 \\ b_2 \\ b_3 \\ b_4 \end{pmatrix} \\
\begin{pmatrix} B_1 \\ B_2 \\ B_3 \\ B_4 \end{pmatrix} &= \begin{pmatrix} b_5 & -b_6 & -b_7 & -b_8 \\ b_6 & b_5 & -b_8 & b_7 \\ b_7 & b_8 & b_5 & -b_6 \\ b_8 & -b_7 & b_6 & b_5 \end{pmatrix} \cdot \begin{pmatrix} a_1 \\ a_2 \\ a_3 \\ a_4 \end{pmatrix}
\end{aligned} \tag{3.67}$$

Plugging in these variables in (3.66) we get

$$\begin{aligned}
&(A_1 + B_1)^2 + (A_2 + B_2)^2 + (A_3 + B_3)^2 + (A_4 + B_4)^2 \\
&\quad - \frac{4(a_8b_5 - a_7b_6 + a_6b_7 - a_5b_8)}{(b_6^2 + b_5^2 + b_7^2 + b_8^2)(a_6^2 + a_7^2 + a_5^2 + a_8^2)} \\
&\quad \times \left[\begin{array}{l} (-b_8B_1 + b_7B_2 - b_6B_3 + b_5B_4)(a_5A_1 + a_6A_2 + a_7A_3 + a_8A_4) \\
\quad + (b_7B_1 + b_8B_2 - b_5B_3 - b_6B_4)(a_6A_1 - a_5A_2 - a_8A_3 + a_7A_4) \\
\quad + (-b_6B_1 + b_5B_2 + b_8B_3 - b_7B_4)(a_7A_1 + a_8A_2 - a_5A_3 - a_6A_4) \\
\quad + (b_5B_1 + b_6B_2 + b_7B_3 + b_8B_4)(a_8A_1 - a_7A_2 + a_6A_3 - a_5A_4) \end{array}\right]
\end{aligned} \tag{3.68}$$

Without loss of generality, we can assume that

$$a_5^2 + a_6^2 + a_7^2 + a_8^2 = 1 \quad \text{and} \quad b_5^2 + b_6^2 + b_7^2 + b_8^2 = 1 \tag{3.69}$$

This, along with the fact that the original expression in (3.66) is homogeneous leads to

$$
\begin{aligned}
&(A_1+B_1)^2+(A_2+B_2)^2+(A_3+B_3)^2+(A_4+B_4)^2 \\
&\quad -4(a_8b_5-a_7b_6+a_6b_7-a_5b_8) \\
&\quad \times \begin{bmatrix} (-b_8B_1+b_7B_2-b_6B_3+b_5B_4)(a_5A_1+a_6A_2+a_7A_3+a_8A_4) \\ \quad +(b_7B_1+b_8B_2-b_5B_3-b_6B_4)(a_6A_1-a_5A_2-a_8A_3+a_7A_4) \\ \quad +(-b_6B_1+b_5B_2+b_8B_3-b_7B_4)(a_7A_1+a_8A_2-a_5A_3-a_6A_4) \\ \quad +(b_5B_1+b_6B_2+b_7B_3+b_8B_4)(a_8A_1-a_7A_2+a_6A_3-a_5A_4) \end{bmatrix}
\end{aligned}
\tag{3.70}
$$

We note that $\det([\mathbf{u}_1|\mathbf{u}_2|\mathbf{u}_3|\mathbf{u}_4]) = \det([\mathbf{v}_1|\mathbf{v}_2|\mathbf{v}_3|\mathbf{v}_4]) = -1$. In fact $\{\mathbf{u}_1, \mathbf{u}_2, \mathbf{u}_3, \mathbf{u}_4\}$ and $\{\mathbf{v}_1, \mathbf{v}_2, \mathbf{v}_3, \mathbf{v}_4\}$ are two orthonormal bases for $\mathcal{R}^4$. Therefore, for every vector $\mathbf{x} \in \mathcal{R}^4$ we have

$$
\begin{aligned}
\mathbf{x} &= (\mathbf{x}.\mathbf{u}_1)\mathbf{u}_1 + (\mathbf{x}.\mathbf{u}_2)\mathbf{u}_2 + (\mathbf{x}.\mathbf{u}_3)\mathbf{u}_3 + (\mathbf{x}.\mathbf{u}_4)\mathbf{u}_4 \\
\mathbf{x} &= (\mathbf{x}.\mathbf{v}_1)\mathbf{v}_1 + (\mathbf{x}.\mathbf{v}_2)\mathbf{v}_2 + (\mathbf{x}.\mathbf{v}_3)\mathbf{v}_3 + (\mathbf{x}.\mathbf{v}_4)\mathbf{v}_4
\end{aligned}
\tag{3.71}
$$

Therefore, we can write the original expression in (3.66) in the form of

$$
\begin{aligned}
&(\mathbf{A}+\mathbf{B})\cdot(\mathbf{A}+\mathbf{B}) - 4(\mathbf{u}_4\cdot\mathbf{v}_4)[(\mathbf{B}\cdot\mathbf{u}_1)(\mathbf{A}\cdot\mathbf{v}_1) + (\mathbf{B}\cdot\mathbf{u}_2)(\mathbf{A}.\mathbf{v}_2) \\
&\quad + (\mathbf{B}\cdot\mathbf{u}_3)(\mathbf{A}\cdot\mathbf{v}_2) + (\mathbf{B}\cdot\mathbf{u}_3)(\mathbf{A}\cdot\mathbf{v}_3) + (\mathbf{B}\cdot\mathbf{u}_4)(\mathbf{A}\cdot\mathbf{v}_4)]
\end{aligned}
\tag{3.72}
$$

where $\mathbf{A} = [\mathbf{A}_1\ \mathbf{A}_2\ \mathbf{A}_3\ \mathbf{A}_4]^T$ and $\mathbf{B} = [\mathbf{B}_1\ \mathbf{B}_2\ \mathbf{B}_3\ \mathbf{B}_4]^T$. Note that

$$
\begin{aligned}
&|\mathbf{u}_1| = |\mathbf{u}_2| = \mathbf{u}_3| = |\mathbf{u}_4| = |\mathbf{v}_1| = |\mathbf{v}_2| = |\mathbf{v}_3| = |\mathbf{v}_4| = 1 \\
&\mathbf{u}_1\cdot\mathbf{v}_1 = \mathbf{u}_2\cdot\mathbf{v}_2 = \mathbf{u}_3\cdot\mathbf{v}_3 = \mathbf{u}_4\cdot\mathbf{v}_4 \\
&\mathbf{u}_1\cdot\mathbf{v}_2 = -\mathbf{u}_2\cdot\mathbf{v}_1 = -\mathbf{u}_3\cdot\mathbf{v}_4 = \mathbf{u}_4\cdot\mathbf{v}_3 \\
&\mathbf{u}_1\cdot\mathbf{v}_3 = -\mathbf{u}_3\cdot\mathbf{v}_1 = \mathbf{u}_2\cdot\mathbf{v}_4 = -\mathbf{u}_4\cdot\mathbf{v}_2 \\
&\mathbf{u}_1\cdot\mathbf{v}_4 = -\mathbf{u}_4\cdot\mathbf{v}_1 = -\mathbf{u}_2\cdot\mathbf{v}_3 = \mathbf{u}_3\cdot\mathbf{v}_2
\end{aligned}
\tag{3.73}
$$

Let us assume $\mathbf{A} = \alpha_1\mathbf{u}_1 + \alpha_2\mathbf{u}_2 + \alpha_3\mathbf{u}_3 + \alpha_4\mathbf{u}_4$ and $\mathbf{B} = \beta_1\mathbf{u}_1 + \beta_2\mathbf{u}_2 + \beta_3\mathbf{u}_3 + \beta_4\mathbf{u}_4$. Also, we define the auxilary vector $\mathbf{C} = \alpha_1\mathbf{v}_1 + \alpha_2\mathbf{v}_2 + \alpha_3\mathbf{v}_3 + \alpha_4\mathbf{v}_4$. Therefore, the expression in (3.72) can be rewritten as

$$
(\mathbf{A}+\mathbf{B})^2 - 4(\mathbf{u}_1\cdot\mathbf{v}_1)\mathbf{C}\cdot\mathbf{B}
\tag{3.74}
$$

We also note that

$$\begin{aligned}\mathbf{A}\cdot\mathbf{C} &= \left(\sum_j \alpha_j \mathbf{u}_j\right)\cdot\left(\sum_i \alpha_i \mathbf{v}_i\right) = \sum_j\sum_i \alpha_j\alpha_i \mathbf{u}_j\cdot\mathbf{v}_i \\ &= \sum_i \alpha_i^2 \mathbf{u}_i\cdot\mathbf{v}_i + \sum_{i<j}\alpha_i\alpha_j(\mathbf{u}_j\cdot\mathbf{v}_i + \mathbf{u}_i\cdot\mathbf{v}_j) \\ &= (\mathbf{u}_1\cdot\mathbf{v}_1)\sum_i \alpha_i^2 \end{aligned} \tag{3.75}$$

Using this we can further simplify the expression in (3.74)

$$\begin{aligned}(\mathbf{A}+\mathbf{B})^2 - 4\frac{\mathbf{A}\cdot\mathbf{C}}{|\mathbf{C}|^2}\mathbf{C}\cdot\mathbf{B} &= |\mathbf{A}|^2 + |\mathbf{B}|^2 + 2\mathbf{A}\cdot\left(\mathbf{B} - 2\frac{\mathbf{C}\cdot\mathbf{B}}{|\mathbf{C}|^2}\mathbf{C}\right) \\ &= |\mathbf{A}|^2 + \left|\mathbf{B} - 2\frac{\mathbf{C}\cdot\mathbf{B}}{|\mathbf{C}|^2}\mathbf{C}\right|^2 + 2\mathbf{A}\cdot\left(\mathbf{B} - 2\frac{\mathbf{C}\cdot\mathbf{B}}{|\mathbf{C}|^2}\mathbf{C}\right) \\ &= \left(\mathbf{A}+\mathbf{B} - 2\frac{\mathbf{C}\cdot\mathbf{B}}{|\mathbf{C}|^2}\mathbf{C}\right)^2 \end{aligned} \tag{3.76}$$

Plugging in a_i and b_i in the last term results in

$$\begin{aligned}&\|\mathbf{H}\|^2\|\mathbf{G}\|^2 - \|\mathbf{H}^\dagger\mathbf{G}\|^2 \\ &\quad= \left(\mathbf{A}+\mathbf{B} - 2\frac{\mathbf{C}\cdot\mathbf{B}}{|\mathbf{C}|^2}\mathbf{C}\right)^2 \\ &\quad= \sum_i\left(A_i + \left[\mathbf{B} - 2\frac{\mathbf{C}\cdot\mathbf{B}}{|\mathbf{C}|^2}\mathbf{C}\right]_i\right)^2 \\ &\quad= \left(a_5b_1 - a_6b_2 - a_7b_3 - a_8b_4 + a_1b_5 + a_2b_6 + a_3b_7 + a_4b_8\right. \\ &\qquad\left. - \frac{2(a_1b_4 + a_3b_2 + a_4b_1 - a_2b_3)(b_1b_8 + b_2b_7 - b_3b_6 + b_4b_5)}{b_1^2 + b_3^2 + b_4^2 + b_2^2}\right)^2 \\ &\qquad + \left(a_6b_1 + a_5b_2 - a_8b_3 + a_7b_4 + a_1b_6 - a_2b_5 + a_3b_8 - a_4b_7\right. \\ &\qquad\left. - \frac{2(a_1b_4 + a_3b_2 + a_4b_1 - a_2b_3)(-b_1b_7 + b_2b_8 + b_3b_5 + b_4b_6)}{b_1^2 + b_2^2 + b_3^2 + b_4^2}\right)^2 \\ &\qquad + \left(a_7b_1 + a_8b_2 + a_5b_3 - a_6b_4 + a_1b_7 - a_2b_8 - a_3b_5 + a_4b_6\right. \\ &\qquad\left. + \frac{2(a_1b_4 + a_3b_2 + a_4b_1 - a_2b_3)(-b_1b_6 + b_2b_5 - b_3b_8 - b_4b_7)}{b_1^2 + b_2^2 + b_3^2 + b_4^2}\right)^2\end{aligned}$$

$$+\Bigg(a_8b_1 - a_7b_2 + a_6b_3 + a_5b_4 + a_1b_8 + a_2b_7 - a_3b_6 - a_4b_5$$
$$+\frac{2(a_1b_4 + a_3b_2 + a_4b_1 - a_2b_3)(b_1b_5 + b_2b_6 + b_3b_7 - b_4b_8)}{b_1 + b_2^2 + b_3^2 + b_4^2}\Bigg)^2 \tag{3.77}$$

which proves the lemma. □

Proof of Lemma 2 Instantaneous SNR equals $|\alpha|^2.|\beta|^2$. Therefore, using (3.3) we have

$$\begin{aligned}
\text{Diversity} &= \lim_{\epsilon\to 0}\frac{\log Pr\{|\alpha|^2.|\beta|^2<\epsilon\}}{\log\epsilon} = \lim_{\epsilon\to 0}\frac{\int_0^\infty f_\beta(z)Pr\{|\alpha|^2<\frac{\epsilon}{z}\}dz}{\log\epsilon}\\
&= \lim_{\epsilon\to 0}\frac{\log(\int_0^\infty e^{-z}[1-e^{-\epsilon/z}]dz)}{\log\epsilon}\\
&= \lim_{\epsilon\to 0}\frac{\log(\int_0^{\epsilon^{1-\delta}} e^{-z}[1-e^{-\epsilon/z}]dz + \int_{\epsilon^{1-\delta}}^\infty e^{-z}[1-e^{-\epsilon/z}]dz)}{\log\epsilon}\\
&= \lim_{\epsilon\to 0}\frac{\log(B+A)}{\log\epsilon}
\end{aligned} \tag{3.78}$$

Given $\delta > 0$ and small enough ϵ

$$\int_{\epsilon^{1-\delta}}^\infty e^{-z}\frac{\epsilon^{1+\delta}}{z}dz \le A \le \int_{\epsilon^{1-\delta}}^\infty e^{-z}\frac{\epsilon}{z}dz \tag{3.79}$$

Also,

$$\begin{aligned}
.5e^{-2} &= \int_1^2 \frac{e^{-2}}{2}dz \le \int_1^2 \frac{e^{-z}}{z}dz \le \int_{\epsilon^{1-\delta}}^\infty \frac{e^{-z}}{z}dz\\
&= \int_{\epsilon^{1-\delta}}^1 + \int_1^\infty \frac{e^{-z}}{z}dz \le \int_{\epsilon^{1-\delta}}^1 \frac{1}{z^{1+\delta}} + \int_1^\infty \frac{1}{z^2}dz = \frac{\epsilon^{-\delta(1-\delta)}-1}{\delta}+1
\end{aligned} \tag{3.80}$$

Therefore,

$$\frac{\epsilon^{1+\delta}}{2e^2} \le A \le \frac{\epsilon^{1-\delta+\delta^2}-\epsilon}{\delta} - \epsilon \tag{3.81}$$

In addition,

$$\epsilon^{1+\delta}.\frac{1}{2}.\epsilon^{1-\delta^2} \le \epsilon^{1+\delta}.e^{-\epsilon^{1-\delta}}[1-e^{-\epsilon^{1-\delta}}] \le B \le [\epsilon^{1-\delta}-0^+].e^{0^+}.[1-e^{-\frac{\epsilon}{0^+}}] = \epsilon^{1-\delta} \tag{3.82}$$

The left-most inequality in the above equation is correct for $0 < \delta < 1$ and small enough ϵ for which $e^{-\epsilon^{1+\delta}} \le \frac{1}{2}$ and $\epsilon^{1-\delta^2} \le [1 - e^{-\epsilon^{1-\delta}}]$ as used in (3.81). Therefore,

$$1 + \delta \le \lim_{\epsilon \to 0} \frac{\log(B + A)}{\log \epsilon} \le 1 - \delta \tag{3.83}$$

This is true for all $0 < \delta < 1$, therefore taking the limit gives us

$$\text{Diversity} = \lim_{\delta \to 0} (1 - \delta) = \lim_{\delta \to 0} (1 + \delta) = 1 \tag{3.84}$$

□

Extention of Lemma 2 *When channel H equals $\sqrt{\sum_{i=1}^{M} |\alpha_i|^2 . \sum_{i=1}^{M} |\beta_i|^2}$, with α_i and β_i being i.i.d. complex Gaussian random variables, the diversity order is M.*

This can be proved using the same technique as above.

Proof of Lemma 3 Plugging in $\mathbf{C}$ by

$$\mathbf{C} = \begin{pmatrix} \mathbf{I} & \mathbf{B}_1\mathbf{B}_2^T & \cdots & \mathbf{B}_1\mathbf{B}_{M-1}^T \\ \mathbf{B}_2\mathbf{B}_1^T & \mathbf{I} & \cdots & \mathbf{B}_2\mathbf{B}_{M-1}^T \\ \vdots & \vdots & \ddots & \vdots \\ \mathbf{B}_{M-1}\mathbf{B}_1^T & \mathbf{B}_{M-1}\mathbf{B}_2^T & \cdots & \mathbf{I} \end{pmatrix} \tag{3.85}$$

we get

$$\mathbf{C} \cdot \begin{pmatrix} a_1\mathbf{B}_1 \\ a_2\mathbf{B}_2 \\ \vdots \\ a_M\mathbf{B}_{M-1} \end{pmatrix} = \begin{pmatrix} (a_1 + a_2\beta_2 + \cdots + a_{M-1}\beta_{M-1})\mathbf{B}_1^T \\ (a_1\beta_1 + a_2 + \cdots + a_{M-1}\beta_{M-1})\mathbf{B}_2^T \\ \vdots \\ (a_1\beta_1 + a_2\beta_2 + \cdots + a_{M-1})\mathbf{B}_{M-1}^T \end{pmatrix} \tag{3.86}$$

and solving for a_i and λ we get

$$\text{for } i = 1, 2, \ldots, M - 1 \tag{3.87}$$

We can always normalize a_i coefficients such that $\sum_{i=1}^{M-1} a_i \beta_i = 1$. Therefore

$$\begin{aligned} a_i &= \frac{1}{\lambda + \beta_i - 1} \quad \text{and} \\ \sum_{i=1}^{M-1} a_i \beta_i &= \sum_{i=1}^{M-1} \frac{\beta_i}{\lambda + \beta_i - 1} = 1 \end{aligned} \tag{3.88}$$

which proves the lemma. □

Proof of Lemma 4 It is clear why none of the roots can be zero. Because, if it is so we will have $\sum_{i=1}^{M} \frac{\beta_i}{\beta_i - 1} = 1$ which is impossible since $\beta_i - 1 < 0$ and $\beta_i > 0$ by definition. Also, from the definition we know β_is are distinct. Therefore, without loss of generality we can assume $\beta_1 < \cdots < \beta_{M-1}$. It will then be easy to show that $f(\lambda) = \sum_{i=1}^{M-1} \frac{\beta_i}{\lambda + \beta_i - 1}$ is monotonic over the following $M - 1$ intervals

$$(1 - \beta_{M-1}, 1 - \beta_{M-2}), \ldots, (1 - \beta_2, 1 - \beta_1), (1 - \beta_1, \infty) \tag{3.89}$$

For the first $M - 2$ intervals, $f(\lambda)$ takes all the values from $-\infty$ to $+\infty$. For the last interval, it takes ∞ at the proximity of $1 - \beta_1$ and 0 when λ goes to ∞. Therefore, it takes the value of 1 in all of these $M - 1$ intervals exactly once, which proves the lemma. □

Appendix B

In this section we briefly review maximum likelihood decoding of an Alamouti code after interference cancellation. As explained in the chapter, when having more than 2 receive antennas, noise terms will be correlated. Therefore, one will need noise-whitening in order to optimally decode the signal. We will show, that this operation will not affect the simple decodablity of Alamouti as follows.

As shown earlier, after canceling the first user we have

$$\begin{aligned} \mathbf{r}'_1 &= \left(\frac{\mathbf{G}_2^\dagger \mathbf{H}_2}{\|\mathbf{G}_2\|^2} - \frac{\mathbf{G}_1^\dagger \mathbf{H}_1}{\|\mathbf{G}_1\|^2}\right)\mathbf{c} + \mathbf{n}'_1 = \mathbf{H}'_1\mathbf{c} + \mathbf{n}'_1 \\ \mathbf{r}'_2 &= \left(\frac{\mathbf{G}_3^\dagger \mathbf{H}_3}{\|\mathbf{G}_3\|^2} - \frac{\mathbf{G}_1^\dagger \mathbf{H}_1}{\|\mathbf{G}_1\|^2}\right)\mathbf{c} + \mathbf{n}'_2 = \mathbf{H}'_2\mathbf{c} + \mathbf{n}'_2 \end{aligned} \tag{3.90}$$

where the correlation matrix of the noise is

$$\mathbf{C}_n = \begin{pmatrix} \left(\frac{\sigma^2}{\|\mathbf{G}_2\|^2} + \frac{\sigma^2}{\|\mathbf{G}_1\|^2}\right)\mathbf{I}_2 & \frac{\sigma^2}{\|\mathbf{G}_1\|^2}\mathbf{I}_2 \\ \frac{\sigma^2}{\|\mathbf{G}_1\|^2}\mathbf{I}_2 & \left(\frac{\sigma^2}{\|\mathbf{G}_3\|^2} + \frac{\sigma^2}{\|\mathbf{G}_1\|^2}\right)\mathbf{I}_2 \end{pmatrix} \tag{3.91}$$

Let us define

$$\begin{aligned} \mathbf{r} &= (\mathbf{r}_1\ \mathbf{r}_2)^T \\ \mathbf{H} &= (\mathbf{H}'_1\ \mathbf{H}'_2)^T \end{aligned} \tag{3.92}$$

The maximum likelihood decoding metric will the be

$$\arg\min_{\mathbf{c}} (\mathbf{r} - \mathbf{H}\mathbf{c})^\dagger \mathbf{C}_\mathbf{n}^{-1} (\mathbf{r} - \mathbf{H}\mathbf{c}). \tag{3.93}$$

It can be shown [36] that

$$\mathbf{C}_n^{-1} = \begin{pmatrix} x\mathbf{I} & y\mathbf{I} \\ y\mathbf{I} & t\mathbf{I} \end{pmatrix} \tag{3.94}$$

Then, the ML criterion will be

$$
\begin{aligned}
&\left(\mathbf{r}_1^{\dagger} - \mathbf{c}^{\dagger}\mathbf{H}_1^{\dagger}\mathbf{r}_2^{\dagger} - \mathbf{c}^{\dagger}\mathbf{H}_2^{\dagger}\right)\begin{pmatrix} x\mathbf{I} & y\mathbf{I} \\ y\mathbf{I} & t\mathbf{I} \end{pmatrix}\begin{pmatrix} \mathbf{r}_1 - \mathbf{H}_1\mathbf{c} \\ \mathbf{r}_2 - \mathbf{H}_2\mathbf{c} \end{pmatrix} \\
&\quad = x\|\mathbf{r}_1 - \mathbf{H}_1\mathbf{c}\|^2 + t\|\mathbf{r}_2 - \mathbf{H}_2\mathbf{c}\|^2 + 2yRe\left\{\left(\mathbf{r}_1^{\dagger} - \mathbf{c}^{\dagger}\mathbf{H}_1^{\dagger}\right)(\mathbf{r}_2 - \mathbf{H}_2\mathbf{c})\right\} \\
&\quad = x\|\mathbf{r}_1 - \mathbf{H}_1\mathbf{c}\|^2 + t\|\mathbf{r}_2 - \mathbf{H}_2\mathbf{c}\|^2 + 2yRe\left\{\mathbf{r}_1^{\dagger}\mathbf{r}_2 - \mathbf{r}_1^{\dagger}\mathbf{H}_2\mathbf{c} - \mathbf{c}^{\dagger}\mathbf{H}_1^{\dagger}\mathbf{r}_2 + \mathbf{c}^{\dagger}\mathbf{H}_1^{\dagger}\mathbf{H}_2\mathbf{c}\right\}
\end{aligned}
\tag{3.95}
$$

The only part in the above equation that could generate cross-terms and therefore cause non-separate decoding is $\mathbf{c}^{\dagger}\mathbf{H}_1^{\dagger}\mathbf{H}_2\mathbf{c}$. Before we expand this term, we note that $\mathbf{H}_1^{\dagger}\mathbf{H}_2$ is in the form of an Alamouti and can be written like $\left(\begin{smallmatrix} h_1 & h_2 \\ -h_2^* & h_1^* \end{smallmatrix}\right)$. Having that in mind the last term in (3.95) can be written as

$$
\begin{aligned}
&Re\left\{h_1|c_1|^2 + h_1^*|c_2|^2 + h_2c_1^*c_2 - h_2^*c_2^*c_1\right\} \\
&\quad = Re\{h_1\}\left(|c_1|^2 + |c_2|^2\right) + 2Re\left\{j \cdot Im\{h_2c_1^*c_2\}\right\} \\
&\quad = Re\{h_1\}\left(|c_1|^2 + |c_2|^2\right)
\end{aligned}
\tag{3.96}
$$

which clearly does not have any cross-terms and therefore c_1 and c_2 can be decoded separately.

Chapter 4
Global Optimal Routing, Scheduling and Power Control for Multi-Hop Wireless Networks with Interference

It happens often that the physical layer algorithm in use is not capable of removing the interference. It then will be with the MAC layer on how to optimize the transmission in order to use less resources of the network, e.g. power. In this chapter we assume full interference from all links in the network with their corresponding weight. After approximating the capacity formula, we then introduce an efficient joint routing, scheduling and power control algorithm that minimizes the consumes power while providing the end-to-end data flow.

1 Modeling and Problem Formulation

Suppose there are N stationary nodes, labeled by the integers $1, 2, \ldots, N$. A set ε of $L_\varepsilon = |\varepsilon|$ transmission links, among the possible $N(N-1)$ links between nodes, make a network topology. These active links are chosen based on the distance, Signal to Interference Noise Ratio (SINR) or some other connectivity measure [47–50]. For simplicity, we assume in this chapter that two nodes constitute a link if the distance between them is less than a threshold. For a given link $l = (i, j)$, the transmitter node i uses a signal power $P(l)$. The "path gain" from node i to node j is given by $G(i, j)$, and models the effects of signal attenuation due to distance, channel fading and shadowing, as well as antenna gain patterns. We assume that the path gains $G(i, j)$ are constant. The transmitting and receiving nodes of link l are denoted by $T(l)$ and $R(l)$ respectively. The received signal power at node $R(l)$ from the transmitter $T(l)$ thus is given by $P(l)G(T(l), R(l))$. However, signals emanating from other transmitters appear the receiver $R(l)$ as interference, and there is thermal noise as well. The SINR for link l is defined as

$$\gamma(l) = \frac{G(T(l), R(l))P(l)}{\sum_{k \neq l} P(k)G(T(k), R(l)) + n_{R(l)}} \tag{4.1}$$

where n_j is the noise power at node j. Let us assume that the efficient bandwidth of channel l is $W(l)$. Assuming Gaussian noise plus interference, the maximum mutual

S.J. Kazemitabar, *Coping with Interference in Wireless Networks*,
Signals and Communication Technology,
DOI 10.1007/978-90-481-9990-7_4,

information of link l will be $X(l) = W(l).\log_2(1+\gamma(l))$. In a low power regime, the SINR value, $\gamma(l)$, is very small. Therefore, we can use the linear approximation for log function and obtain

$$X(l) = W(l).\gamma(l) \tag{4.2}$$

Scheduling and Power Control

We start with reviewing the problem formulation given in [37]. For simplicity of exposition, we divide time into slots, each of equal duration and indexed by positive integers. Transmissions begin and end on slot boundaries. Generalizing the notation introduced earlier, let $X_t(l)$ and $P_t(l)$ be the data rate for link l in slot t, and transmission power for the transmitter $T(l)$ for link l in slot t, respectively. Let $P_t = (P_t(1), P_t(2), \ldots, P_t(L_\varepsilon))$ be the network power vector for slot t. Let $P^{\max}(i)$ be the maximum transmission power for node i. Also, let $\varepsilon(i)$ be the links in ε that originate at node i. Each node must conform to the peak transmission power constraint in every slot:

$$\begin{aligned} &0 \le \sum_{l\in\varepsilon(i)} P_t(l) \le P^{\max}(i) \quad \text{and} \\ &0 \le P_t(l), \quad \text{for all } t \ge 1 \text{ and } l \in \varepsilon \end{aligned} \tag{4.3}$$

The above constraints form a polytope in $\overrightarrow{P}$ space. Let us denote this polytope by $\mathbb{P}$. Using (4.1) and (4.2), the maximum achievable data rate for link l in slot t is

$$X_t(l) = W(l)\left(\frac{G(T(l), R(l))P_t(l)}{\sum_{k\neq l} G(T(k), R(l))P_t(k) + n_{R(l)}}\right) \tag{4.4}$$

The average rate of link l is then defined as $X_{\text{avg}}(l) = \liminf_{t\to\infty} \frac{1}{t}\sum_{k=1}^{t} X_k(l)$. For each link l, let $C(l)$ be a given minimum required average data rate, i.e. we must have

$$X_{\text{avg}}(l) \ge C(l), \quad \text{for all } l \in \varepsilon \tag{4.5}$$

Define the required minimum average rate vector as $\overrightarrow{C} = (C(1), C(2), \ldots, C(L_\varepsilon))$.

The average power consumed by the transmitter of link l is then $P_{\text{avg}}(l) = \limsup_{t\to\infty} \frac{1}{t}\sum_{k=1}^{t} P_k(l)$. Define the average power vector as $\overrightarrow{P_{\text{avg}}} = (P_{\text{avg}}(1), P_{\text{avg}}(2), \ldots, P_{\text{avg}}(L_\varepsilon))$. There may or may not exist a sequence of network power vectors $P_1, P_2, \ldots$ that satisfy (4.3) and (4.5). If there does exist a sequence of such network power vectors, our aim is to minimize a linear function of P_{avg}. An example of such a linear function is simply the total average power

$$h(P_{\text{avg}}) = \sum_{l\in\varepsilon} P_{\text{avg}}(l) \tag{4.6}$$

The *Scheduling and Power Control Problem* is then defined as

$$\min h(P_{\text{avg}}) \quad \text{subject to} \quad (4.3) \text{ and } (4.5) \tag{4.7}$$

Let us define the value of this cost function as a function of $\overrightarrow{C}$ denoted by $H(\overrightarrow{C})$. We can absorb constraint (4.5) into the cost function and define the potential function

$$V(\overrightarrow{P}, \beta) = h(\overrightarrow{P}) + \sum_{l \in \varepsilon} \beta(l)[C(l) - X(l)] \tag{4.8}$$

Using duality methods, we can show that [37]

$$H(\overrightarrow{C}) = \max_{\beta \geq 0} \left\{ \min_{\overrightarrow{P}} V(\overrightarrow{P_{\text{avg}}}, \beta) \text{ subject to (4.3)} \right\} \tag{4.9}$$

Computation of (4.9) involves optimizing over all schedules of network power vectors satisfying the peak power constraint in every slot. However, since the potential function V is linear in $\overrightarrow{P}, \overrightarrow{C}$, and $\overrightarrow{X}$, it follows that (4.9) can be computed by an optimization over a single slot [37, 38]. Therefore, we can just focus on solving

$$\begin{aligned} &\max_{\beta \geq 0} \left\{ \min_{\overrightarrow{P}} V(\overrightarrow{P}, \beta) \right\} \\ &\text{s.t.} \\ &\overrightarrow{P} \in \mathbb{P} \end{aligned} \tag{4.10}$$

It can be shown [37] that V is element-wise concave in terms of $\overrightarrow{P}$. Therefore, the minimization problem takes its optimal value at the extreme points of polytope $\mathbb{P}$. In other words, the optimal schedule should be in such a way that at most one of the links emanated from each node is active. That link, if there is any, should be sending data at full allowable power [38].

For simplicity, we show the mth extreme point of $\mathbb{P}$ by P^m_{ext}. Considering M as the total number of extreme points, we will need to solve the following max-min problem:

$$H(\overrightarrow{C}) = \max_{\beta \geq 0} \left\{ \min_{m} V\left(P^m_{\text{ext}}, \beta\right) : 1 \leq m \leq M \right\} \tag{4.11}$$

This optimization problem can be solved by a linear program:

$$\begin{aligned} &\max T \\ &\text{s.t.} \\ &T - \sum_{l \in \epsilon} P^m_{\text{ext}}(l) - \sum_{l \in \epsilon} \beta(l)(C(l) - X^m(l)) \leq 0 \quad m = 1, \ldots, M \end{aligned} \tag{4.12}$$

The problem with this method is, M, the number of extreme points (constraints) grows exponentially with the number of links in the network, and therefore it is not possible to find the solution for large values of M. In the next section, we introduce an efficient mechanism to solve this optimization problem.

The number of the extreme (edge) points that minimize $V(\overrightarrow{P}, \beta)$ is in general more than one. Let us denote these extreme points by P^i. If the assigned rate vector $\overrightarrow{C}$ is feasible, an optimal schedule of network power vectors exists that consists solely of these extremal network power vectors P^i. Moreover, since a hyperplane in $(L_\varepsilon + 1)$ dimensional Euclidean space is determined by $(L_\varepsilon + 1)$ linearly independent points contained within it, an optimal policy can be constructed that consists of at most $(L_\varepsilon + 1)$ extremal network power vectors P^i. Let K be the number of extremal network power vectors P^i such that (10) holds. It is easy to see that in any optimal policy (assuming feasibility), the average rate on each link l, $X_{\text{avg}}(l)$, is exactly equal to $C(l)$. Otherwise, we could have transmitted less power and still satisfy the rate constraints. Let $X^i(l)$ denote the rate of link l corresponding the optimal extremal network power vector P^i. Assuming feasibility, it can be shown that there exists a vector $\overrightarrow{\lambda}$ such that [37]:

$$\begin{aligned} &\sum_{i=1}^{K} \lambda_i = 1 \\ &X_{\text{avg}}(l) = \sum_{i=1}^{K} \lambda_i X^i(l) = C(l) \end{aligned} \tag{4.13}$$

The value of λ_i indicates the relative frequency that the power vector P^i should be utilized in an optimal policy. This then determines a family of optimal policies, which can be constructed by alternating between the optimal extremal power vectors in accordance with the weight vector λ_i.

Therefore, the main challenge is to find the optimal solutions of (4.12). In our algorithm, as will be shown in the next section, we insert the constraints iteratively to take advantage of the fact that we need at most $L_\epsilon + 1$ transmission modes in the optimal solution.

Scheduling, Power Control, and Routing

In the power control-scheduling problem link rates, $C(l)$, were fixed and the algorithm finds the optimal power allocation vectors and their scheduling frequency. The routing problem should find the optimal link rates for a given end-to-end traffic demand matrix as well.

Now, assume we have a set of source-destination (S-D) pairs $\{(i, j)\}$. Suppose that for each S-D pair the desired traffic demand is known. Solution to the joint routing, scheduling and power control problem specifies links' average traffic rates, power vectors and scheduling frequency of power vectors such that the given traffic demand and maximum power constraints are satisfied. Our objective is to find the solution with minimum average power consumption. Let us then define $v_{ij}(l)$ as the portion of flow in link l that belongs to the S-D pair $\{(i, j)\}$. The mathematical statement of the joint routing scheduling, and power control problem will be as follows,

$$
\begin{aligned}
&\min \sum_{l} P_{\text{avg}}(l) \\
&\text{s.t.} \\
&\text{a.} \quad 0 \leq \sum_{l \in \varepsilon(i)} P_t(l) \leq P^{\max}(i) \quad \text{and} \\
&\text{b.} \quad 0 \leq P_t(l), \quad \text{for all } t \geq 1 \text{ and } l \in \epsilon \\
&\text{c.} \quad \sum_{l \in \mathcal{I}_p} v_{ij}(l) - \sum_{l \in \mathcal{O}_p} v_{ij}(l) = \begin{cases} 0 & p \neq i \text{ or } j \\ -d_{ij} & n = i \\ d_{ij} & n = j \end{cases} \\
&\text{d.} \quad \sum_{ij} v_{ij}(l) = C(l) \\
&\text{e.} \quad X_{\text{avg}}(l) \geq C(l), \quad \text{for all } l \in \epsilon
\end{aligned}
\tag{4.14}
$$

where $\mathcal{I}_n$ and $\mathcal{O}_n$ are sets of links entering and leaving node n respectively. Equations (4.14c) and (4.14d) define a polytope that we denote by $\mathbb{C}$. For a rate vector $\overrightarrow{C}$ to be a route, it must be on this polytope. In [37], a gradient projection algorithm is proposed to find the routing, i.e. the power-optimal rate vector $\overrightarrow{C}$. At each iteration for the given rate vector $\overrightarrow{C}$ on $\mathbb{C}$, the scheduling and power control problem described in Sect. II.A is solved. Then, the gradient projection method is used to update vector $\overrightarrow{C}$ while keeping it on $\mathbb{C}$. Note that the gradient directions are given by the dual variables β [38].

This algorithm is not scalable for large networks for two reasons. First, recall that the proposed solution for scheduling and power control needs to consider all extreme points of the power polytope $\mathbb{P}$, which grows exponentially with number of links. Second, we have to iteratively solve the scheduling and power control problem after each update of the rate vector using gradient projection method, which itself is very time-consuming. It is needless to mention that it will use up a lot of memory to store all the constraints for the linear program of the scheduling algorithm. The algorithm then, will not be efficient for networks with more than 15 links [38].

2 Power Control, Scheduling and Routing Algorithm

In this section we will introduce a new algorithm that can find the optimal power control, scheduling, and routing for large networks. We first explain the power control-scheduling algorithm for the fixed routing and then extend the algorithm to compute the optimal power control, scheduling, and routing simultaneously.

The Power Control-Scheduling Algorithm

Recall that to compute the optimal scheduling and power control for a given routing we have to solve the optimization problem given in (4.12). The problem with this method is that the number of extreme points (constraints), M, grows exponentially with the number of links in the network, and therefore it is not possible to find the solution for large values of M.

A common approach to deal with large number of constraints in linear programs is cutting plane methods [51].[1] Instead of dealing with all constraints, cutting plane algorithm considers a subset of constraints I and form the relaxed problem:

$$\begin{aligned} &\max T \\ &\text{s.t.} \\ &T - \sum_{l \in \epsilon} P^m(l) - \sum_{\substack{l \in \epsilon \\ m \in I}} \beta(l)(C(l) - X^m(l)) \leq 0 \end{aligned} \tag{4.15}$$

We use Kelly's convex cutting plane algorithm [51] which is basically an iterative algorithm for introducing new constraints into the constraint subset. We initialize the constraint set I_1 by selecting one of the extreme points of $\mathbb{P}$ arbitrarily and forming its corresponding constraint.

At each iteration, e.g. k, we solve the relaxed optimization problem (4.15) with constraint set I_k instead of I. Let (T_k, β_k) be the optimal solution of the relaxed problem. There are two possibilities: (1) If (T_k, β_k) is a feasible solution for the original optimization problem, then we are done and we have found an optimal solution for that problem too. (2) If (T_k, β_k) is not feasible we have to find a violated constraint and add it to the constraint set to form I_{k+1} and start the next iteration. We need a method to check if (T_k, β_k) is a feasible solution.

Definition Suppose that we have a candidate solution for an LP. A separation *Oracle* determines if the solution is feasible. If the solution is not feasible, the separation Oracle finds a hyperplane that separates the solution from the feasible region.

We formulate the following Oracle optimization problem:

$$\begin{aligned} &\min\left(\sum_{l} P(l) + \beta_k(l)(C(l) - X(P(l)))\right) \\ &\text{s.t.} \\ &\vec{P} \in \mathbb{P} \end{aligned} \tag{4.16}$$

As explained before, subject to feasibility, there is always an optimal solution in the extreme points of the power polytope, $\mathbb{P}$. Therefore, similar to the concave programming problems, branch and bound algorithms converge in finite number of iterations, and in fact, in most practical cases the algorithm finds the optimal solution

[1] In fact, cutting plane methods are usually used in the more general context of convex programming.

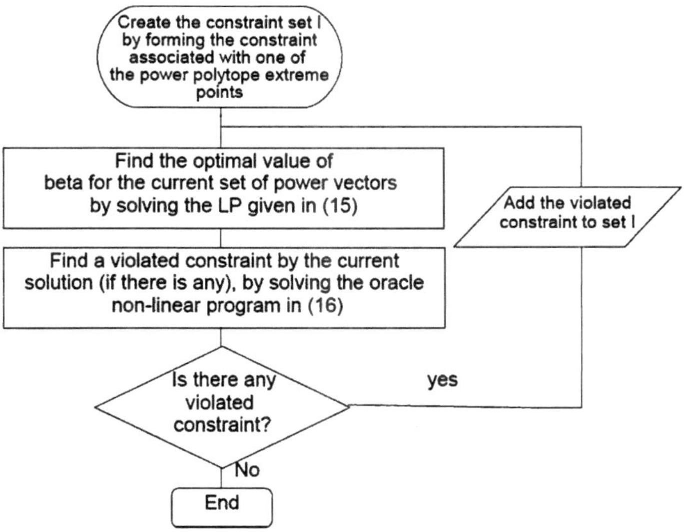

Fig. 4.1 Flowchart of the power control-scheduling algorithm

fast. We use GAMS [52] software which is based on the branch and bound algorithm. The library of GAMS that analytically finds the global optimum is BARON [53]. BARON uses a modified branch and bound method named as branch and reduce. The linear program as well as the core code is performed by MATLAB software. We link the two softwares via MATGAMS, a free interface for GAMS and MATLAB. In the simulation results we elaborate more on the computation time of the algorithm. The essence of the algorithm is as follows.

Figure 4.1 shows the flow chart of the power control-scheduling algorithm. In summary, at iteration k, we solve the following linear program:

$$\begin{aligned}&\max T\\&T-\sum_{l\in\epsilon}P^m(l)-\sum_{l\in\epsilon}\beta(l)(C^m(l)-X^m(l))\le 0\\&m=1,\dots,k\end{aligned}\tag{4.17}$$

Let β_k be the solution of the above LP in iteration k, which will be used in the following non-linear Oracle problem to test optimality:

$$\begin{aligned}&\min_{\vec{P}}\left(\sum_{l}P(l)+\sum_{l}\beta_k(l)(C(l)-X(l))\right)\\&\text{s.t.}\\&\vec{P}\in\mathbb{P}\end{aligned}\tag{4.18}$$

We then add the new $\overrightarrow{P}$ vector to form a new constraint for the next iteration. In the following we prove that the algorithm converges in finite iterations to the optimal solution.

Lemma 1 *The proposed algorithm converges after finite number of iterations.*

Proof At the end of each iteration, if the convergence criterion is not satisfied, it means that the Oracle has found a new power vector that its corresponding constraint is violated. Since solutions of the Oracle are extreme points of the power polytope, and there are finite number of them, it proves that the algorithm converges in finite iterations. □

In fact, in our simulations, algorithm converges after around L_ϵ iterations.

Lemma 2 *After convergence the value of both linear and nonlinear (global) optimizers will be equal to* $H(\overrightarrow{C})$.

Proof Let us denote the total number of extreme points by M. At the kth iteration we will have k power modes (extreme points). Let us define

$$g^k(\beta) = \min_{\overrightarrow{P}^1,\ldots,\overrightarrow{P}^k} V(\overrightarrow{P},\beta) \tag{4.19}$$

As was shown for LP in (4.14), it is easily seen the value of the linear program at Iteration k, T^k, will be equal to $\max_{\beta\geq 0} g^k(\beta)$. We know the global minimum of $V(P,\beta)$ is on one of the extreme points. Therefore, the solution of the Oracle problem defined in (4.16) at the kth iteration is

$$\min_{\overrightarrow{P^1},\ldots,\overrightarrow{P^M}} V(\overrightarrow{P},\beta_k) = g^M(\beta_k) \tag{4.20}$$

where β_k is the optimal β vector found by the linear program (15) in Iteration k.

By definition, $g^k(\beta) \geq g^M(\beta)$. Therefore,

$$g^k(\beta_k) = \max_{\beta\geq 0} g^k(\beta) \geq \max_{\beta\geq 0} g^M(\beta) = H(\overrightarrow{C}) \tag{4.21}$$

This means $\min_{\overrightarrow{P^1},\ldots,\overrightarrow{P^k}} V(\overrightarrow{P},\beta_k) \geq \min_{\overrightarrow{P^1},\ldots,\overrightarrow{P^M}} V(\overrightarrow{P},\beta_M)$.

Suppose the algorithm converges after m iterations. The convergence criterion suggests the value of nonlinear program being more than or equal to the value of linear program:

$$\min_{\overrightarrow{P^1},\ldots,\overrightarrow{P^M}} V(\overrightarrow{P},\beta_m) \geq \max_{\beta\geq 0} \min_{\overrightarrow{P^1},\ldots,\overrightarrow{P^m}} V(\overrightarrow{P},\beta) = \min_{\overrightarrow{P^1},\ldots,\overrightarrow{P^M}} V(\overrightarrow{P},\beta_k) \tag{4.22}$$

From that, we conclude that

$$g^M(\beta_M) = \max_{\beta\geq 0} g^M(\beta) \geq g^M(\beta_m) \geq g^m(\beta_m) \tag{4.23}$$

This along with (4.21) results in

$$g^M(\beta_M) = H(\overrightarrow{C}) = g^m(\beta_m) \tag{4.24}$$

which proves the lemma. □

Theorem 1 *The set of power modes after convergence, $\{\overrightarrow{P^1}, \ldots, \overrightarrow{P^m}\}$, satisfy the rate requirement. In other words, there is a non-negative time sharing, $\lambda_1, \ldots, \lambda_m$ that*

$$\begin{aligned} &\text{a.} \quad \sum \lambda_i = 1 \\ &\text{b.} \quad \sum \lambda_i . \overrightarrow{X}^i = \overrightarrow{C} \\ &\text{c.} \quad \sum \lambda_i \sum_l P^i(l) = H(\overrightarrow{C}) \end{aligned} \tag{4.25}$$

Proof Let us rewrite the linear program in the iteration that leads to convergence. We use the standard form of convex optimization, i.e. cost-constraint format. We denote the cost function of the LP problem by f and the constraint functions by h_i [54].

$$\begin{aligned} f(T, \beta(1), \ldots, \beta(L)) &= T \\ h_1(T, \beta(1), \ldots, \beta(L)) &= T - \sum_l \beta(l)(C(l) - X^1(l)) - \sum_l P^1(l) \leq 0 \\ &\vdots \\ h_m(T, \beta(1), \ldots, \beta(L)) &= T - \sum_l \beta(l)(C(l) - X^m(l)) - \sum_l P^m(l) \leq 0 \end{aligned} \tag{4.26}$$

Since the linear program finds the global minimum of the convex (linear) cost function, K.K.T. conditions are satisfied.

$$\nabla f = \sum_{i=1}^{m} \lambda_i \nabla h_i \tag{4.27}$$

The above vector equation gives us $L_\epsilon + 1$ scalar equations. The first one is $\sum_i \lambda_i = 1$, and the rest are

$$\sum_i \lambda_i X^i(l) = C(l) \quad \text{for } l = 1, \ldots, L_\epsilon \tag{4.28}$$

This proves (25a) and (25b). Then, as Lemma 2 suggests, after convergence $T = H(\overrightarrow{C})$. One of the conclusions of the K.K.T. theorem is that at the optimal point, either the dual variable, λ, or the constraint, h, should be equal to zero [54]. Using this fact along with (4.28) we get

$$\sum \lambda_i \sum_l P^i(l) = H(\vec{C}) \tag{4.29}$$

□

It can be shown that the λ coefficients are the dual variables of the LP (4.17) and since they satisfy (4.28), they present the frequency of use of each transmission mode. Therefore, we do not need to solve the additional LP given in (4.13) to find the scheduling frequencies.

Power Control, Scheduling and Routing

We extend the power control-scheduling algorithm developed in Sect. III.A to find the optimal routing too. The routing algorithm described in [37] has scalability problems, since it has to consider all extreme points of the power polytope explicitly and relies on iterative gradient based updates of the routing parameters. Both these issues are resolved in our algorithm.

The problem formulation is given in (4.14). In Sect. II.A, we explained that by using duality and due to linearity in P and X, we need to only solve the optimization problem given in (4.10). Similarly, we can use the duality theorem for linear programs and linearity of (4.14) in P, X and C to show that for the joint power control, scheduling, and routing problem it is sufficient to solve the following optimization problem.

$$\begin{aligned} &\max_{\beta\geq 0}\Big\{\min_{\vec{P},\vec{C}} V(\vec{P},\beta,\vec{C})\Big\} \\ &\text{s.t.} \\ &\vec{P}\in\mathbb{P} \quad \text{and} \quad \vec{C}\in\mathbb{C} \end{aligned} \tag{4.30}$$

Here, we introduce $\vec{C}$ as one of the arguments of function $V()$, since it is not constant and we seek to find its optimal value. The problem can be rewritten as an LP:

$$\begin{aligned} &\max T \\ &\text{s.t.} \\ &T - V(\vec{P},\beta,\vec{C}) \leq 0 \quad \text{for all } \vec{P}\in\mathbb{P} \text{ and } \vec{C}\in\mathbb{C} \end{aligned} \tag{4.31}$$

Recall that the function $V()$ is element wise concave in $\vec{P}$ and linear in $\vec{C}$. Hence, the optimal solution can be found in extreme points of polytopes $\mathbb{P}$ and $\mathbb{C}$. Similar

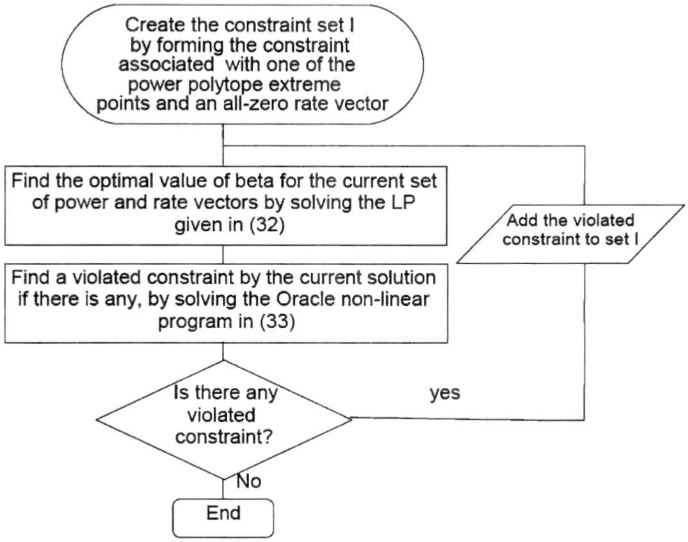

Fig. 4.2 Flowchart of the routing algorithm

to the power control-scheduling case we use the Kelly's cutting plane method to devise an iterative algorithm as follows:

At Iteration k we solve the following master linear program with k constraints:

$$\begin{aligned} &\max T \\ &T - \sum_{l \in \epsilon} P^m(l) - \sum_{l \in \epsilon} \beta(l)(C^m(l) - X^m(l)) \leq 0 \\ &m = 1, \ldots, k \end{aligned} \tag{4.32}$$

Let (T_k, β_k) be the optimal solution of (4.32) at Iteration k; β_k, will be used in the following non-linear Oracle problem

$$\begin{aligned} &\min_{\vec{P}, \vec{C}} \left(\sum_l P(l) + \sum_l \beta_k(l)(C(l) - X(l)) \right) \\ &\text{s.t.} \\ &\vec{C} \in \mathbb{C} \text{ and } \vec{P} \in \mathbb{P} \end{aligned} \tag{4.33}$$

If the Oracle optimal value is less than T_k, then the corresponding solution $(\vec{P}^{k+1}, \vec{C}^{k+1})$ forms a violated constraint that should be added to the master LP constraint set for the next iteration. Otherwise, the algorithm has converged.

Figure 4.2 shows the algorithm explained above.

We can further simplify the last step as follows. The potential function in (4.33) is separable to two functions. The first one, i.e. $\sum_l \beta_k(l)C(l)$, is the sum of weighted $C(l)$s and can be minimized via *Dijkstra* algorithm. The second one is simply the potential function in the previous subsection minus a constant value. Therefore, it does not require a new technique to find the minimum for.

Similar to the scheduling-only case, since the optimal solution is in the extreme points of the power and routing polytopes, the cutting plane algorithm converges in finite number of steps.

After convergence, the optimal routing vector and the optimal transmission strategy will be superposition of all the $\overrightarrow{C}^k$ vectors and $\overrightarrow{C}^k$ vectors respectively. The corresponding coefficients are similar and are found in the same manner as described in the proof of Theorem 1.

3 Nonlinear vs. Linear

Throughout the chapter so far, we have assumed that we can approximate the nonlinear logarithm formula of capacity with its linear equivalent, in the low SNR regime. In this section, we briefly describe the solution for the nonlinear case where we do not approximate. The minimization problem to consider is,

$$
\begin{aligned}
&\min \sum_{l} P_{\text{avg}}(l) \\
&\text{s.t.} \\
&\text{a.}\quad 0 \le \sum_{l\in\varepsilon(i)} P_t(l) \le P^{\max}(i) \quad \text{and} \\
&\text{b.}\quad 0 \le P_t(l), \ \text{for all } t \ge 1 \text{ and } l \in \epsilon \\
&\text{c.}\quad \sum_{l\in\mathcal{I}_p} v_{ij}(l) - \sum_{l\in\mathcal{O}_p} v_{ij}(l) = \begin{cases} 0 & p \ne i \text{ or } j \\ -d_{ij} & n = i \\ d_{ij} & n = j \end{cases} \\
&\text{d.}\quad \sum_{ij} v_{ij}(l) = C(l) \\
&\text{e.}\quad X^{NL}_{\text{avg}}(l) \ge C(l), \quad \text{for all } l \in \epsilon
\end{aligned}
\tag{4.34}
$$

where

$$
\begin{aligned}
X^{NL}_t(l) &= W(l)\log\left(1 + \frac{G(T(l),R(l))P_t(l)}{\sum_{k\ne l} G(T(k),R(l))P_t(k) + n_{R(l)}}\right) \\
X^{NL}_{\text{avg}}(l) &= \lim_{t\to\infty}\inf \frac{1}{t}\sum_{k=1}^{t} X^{NL}_k(l)
\end{aligned}
\tag{4.35}
$$

The structure of the above formulation is similar to that of (4.14) except for nonlinearity of the X function. Also, in the proof of Lemmas 2–3 and Theorem 1 we did not use the linearity of X function (or concavity of the potential function). Therefore, we can use the iterative algorithm in the previous section along with the cutting plane method. After convergence of the algorithm, we can make sure we have reached the globally optimal solution. The only remaining issue is Lemma 1 which means that for this problem, the algorithm is not guaranteed to converge in finite time, due to the nonlinearity of the log function (non-concavity of the new potential function). Since our goal in this section is not optimizing the nonlinear problem and is just comparison with the linear case, the above described algorithm will serve us adequately.

In order to compare the results of the nonlinear, i.e. actual, model with that of the linear, i.e. approximate, model we need to set the traffic demands fairly for the test. In other words, the value of the traffic demand of the two models should be the same. The average value of the final rate in linear model was calculated to be superposition of linear combination of X^m vectors, with $X^m = W.\overrightarrow{SINR^m}$. Therefore, to have a fair comparison, we need the linear combination of $W.\log(\overrightarrow{SINR^m})$s in the linear model to be equal to the traffic demand of the nonlinear model.

As mentioned earlier, the algorithm we proposed for nonlinear model is theoretically not guaranteed to converge in finite time. In real simulations however, they all did converge but took much longer time compared with the linear model.[2] This slowness was due to two facts. First, as the theory predicted, the number of iterations it took the program to run before convergence was increased. Second, the operations in each iteration were computationally more complex because of introducing the logarithm function.

As will be discussed in the next section, the result of the series of simulations showed that the accuracy of the linear model depends on the value of $P^{\max}$. We know that the final solution proposed by the linear model is superposition of some power modes. Each power mode in the linear case is such that only one link per node is at most active, and is transmitting with full power, i.e. $P^{\max}$. Therefore, in order for the model to be precise, we intuitively need $\log(1 + \frac{P^{\max}}{\eta})$ to be close to $\frac{P^{\max}}{\eta}$. This intuition is verified as shown in simulation results of the next section.

4 Simulation Results and Discussion

We consider a square area of 200 by 200 meters for our simulations. The node locations are selected by a uniform random generator in this area. We consider networks with 7 to 50 nodes in our simulations. However, in most of the experiments the number of nodes ranges from 7 to 30. There is an edge between two nodes of the network graph if the distance between them is less than a specific value denoted by $d_{\max}$. The number of edges ranges from 9 to 102 in our experiments, with 9 corresponding to a network with 7 nodes and 102 to a network with 50 nodes. Each edge

[2] In some cases up to 5 times slower.

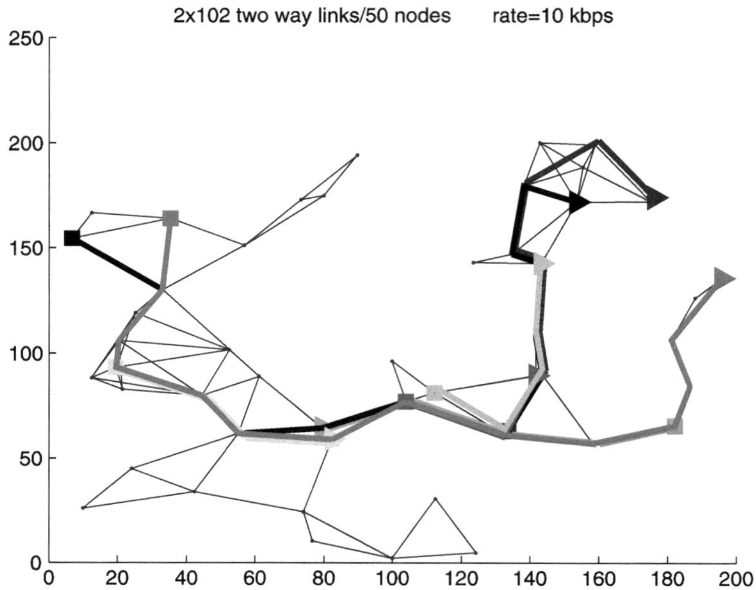

Fig. 4.3 One of the simulated networks with 50 nodes

represents two uni-directional communication links in the network. Therefore, the number of communication links is twice the number of edges of the network graph. For example, in the network of Fig. 4.3 we have 102 edges which is equivalent to 204 uni-directional links. The path-loss and shadowing parameter, $G(i, j)$, is considered proportional to inverse square of the distance between nodes i and j. The algorithm, however, works for any other choice of $G(i, j)$.

The efficient bandwidth of links is 1 MHz, and the noise power .000010 Watt. The maximum power of a node is 1 Watt. For each network, we randomly pick the source-destination pairs. The number of source-destination pairs in a network is about 20% of the total number of nodes. The end-to-end data-rate from each source to its corresponding destination is 10 Kbps, so that in all experiments a feasible solution exists. However, the data rate value does not affect the properties and efficiency of the algorithm. For each network size, multiple random networks were generated and the results were consistent among different runs. Here, we report the average results of at least 3 experiments for each network size. Three separate algorithms are studied here: (1) optimal scheduling and power control using an LP with all extreme points constraint with minimum hop routing, (2) our proposed algorithm for scheduling and power control with minimum hop routing and (3) our proposed joint routing, scheduling and power control algorithm.

In Fig. 4.3 the source and destination pairs and the links that are utilized in the optimal routing solution for each pair are specified. We represent each source node by $\triangleright$ and the destination node by $\square$. The same color is used for these symbols and the links that are in the optimal route between them.

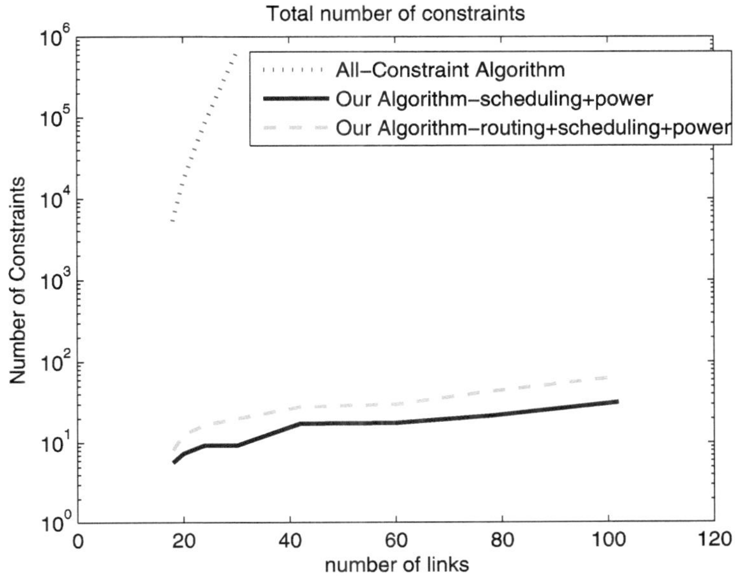

Fig. 4.4 Comparison of the final constraint counts in the optimization LP

Figure 4.4 shows the number of constraints that are in the final version of the optimization problem. In our algorithms, we add one more constraint after each iteration, therefore this is the number of iterations for our algorithm too. The fact that the number of iterations is very close to the number of links, shows that our algorithm is very efficient. Recall that in the optimal solution there are at least $L_{\epsilon}+1$ active constraints, hence there should be at least this number of constraints present in the final LP. We compare this value with the corresponding value in the all-constraint method, which is the total number of extreme points. As was explained in the chapter this number increases exponentially with the number of links in the network.

The number of constraints directly affects the running time of the algorithm. Figure 4.5 compares the running time of the algorithms.[3] We were not able to run the all-constraint method for networks with more than 12 edges. Besides the time, memory is also an issue for the all-constraint method. For example, a 15-edged network has at least $2^{15\times 2}=1$ Gig extreme points that needs to be stored. Both algorithms introduced in this chapter are much more time efficient than the previously proposed algorithms.

If we compare running time curves of the two algorithms introduced here, it becomes clear that finding the optimal routing can be very time consuming. In the following we compare the total power consumption of the optimal routing with the minimum hop routing when we use optimal scheduling and power control in both

[3] All simulations were performed by a 2.6 GHz Compaq Xeon computer, with 256 kb cache and 1 GB of memory.

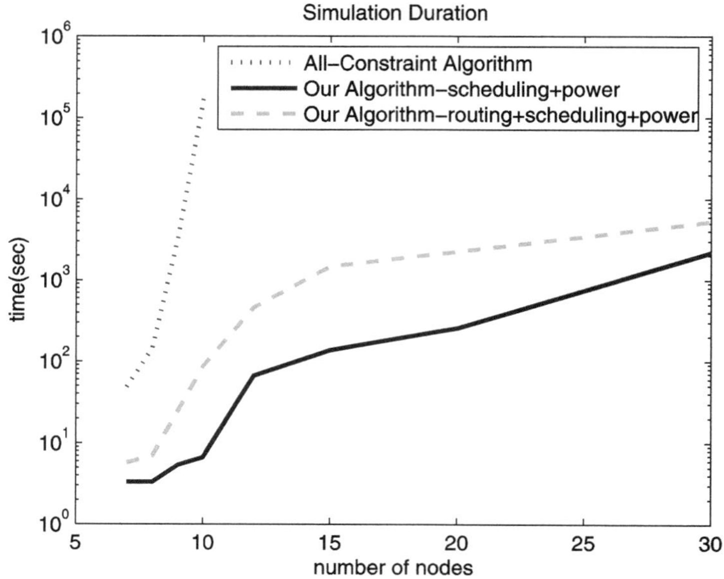

Fig. 4.5 Comparison of the time it takes for each algorithm to converge

cases to quantify the performance gain achieved with optimal routing. In Fig. 4.6 we have plotted the amount of power consumed for two different algorithms. The optimal routing algorithm, on average, consumes around 15 percent less power. This shows that the minimum-hop routing is not power efficient and we can reduce the power consumption by using optimal routes.

In the following, we compare the characteristics of the minimum-hop routing with the optimal routing. Figure 4.7 shows the number of active links in each routing. As can be seen the number of links used by the joint routing and scheduling method is more than that of the minimum-hop method. However, the difference is not significant and it becomes marginal for larger networks. This result may seem to be counter-intuitive since it suggests that the optimal routing would *not* distribute the traffic among the links in the network.

To further study optimal routing characteristics, Fig. 4.8 shows the topology and optimal routes for 3 separate networks with 102 links. The optimal paths tend to use the same links. This observation seems to be in contrary to the previous results in distributed algorithms [42, 43]. There are two issues that should be considered here: (i) For those algorithms that are based on the notion of back-pressure scheduling [43], the proposed distributed algorithms are asymptotically optimal so they may very slowly converge to the optimal paths suggested by our algorithm. However their transient behavior can be very poor since they start with distributing the traffic among many links. (ii) It is very important to understand and distinguish the interference model that is considered for different algorithms. For instance, in [42] it is assumed that there is no interference between two links that do not have any com-

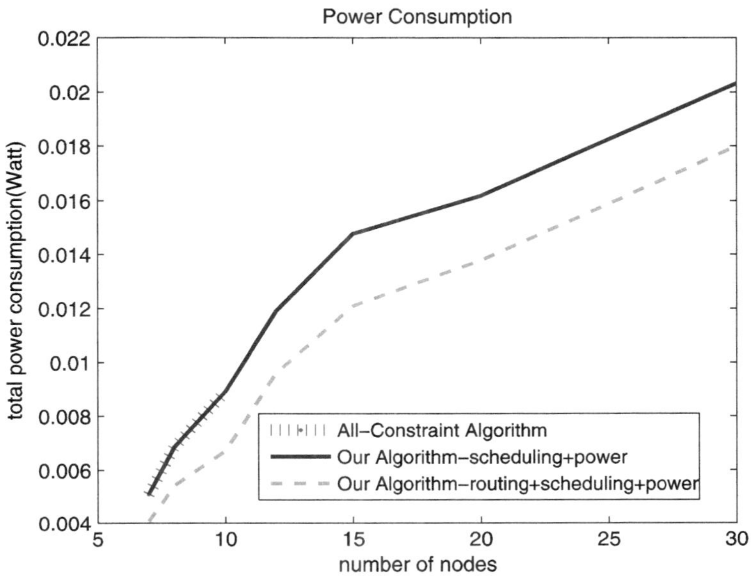

Fig. 4.6 Comparison of the total consumed power

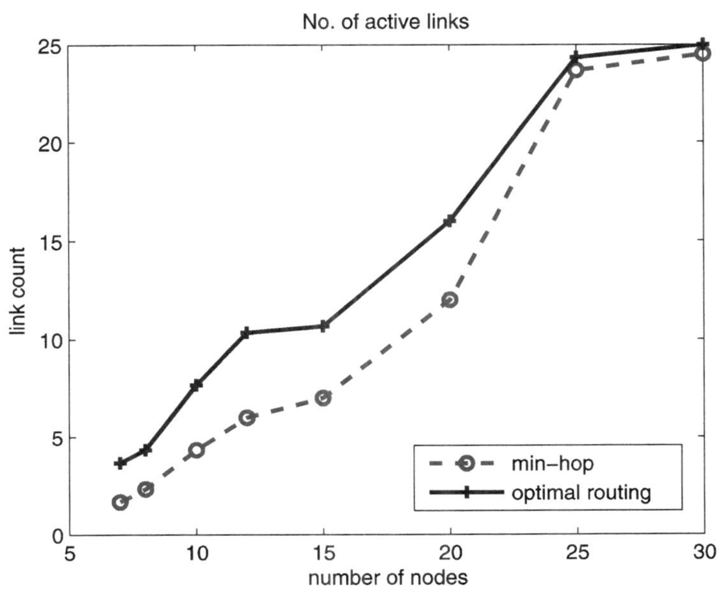

Fig. 4.7 Comparison of the number of active links

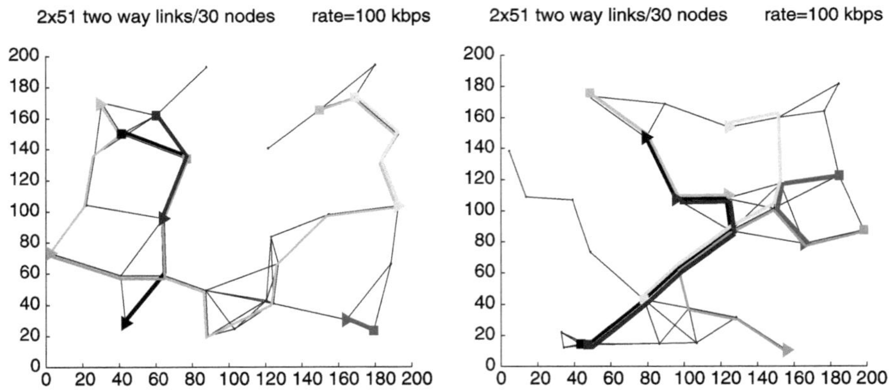

Fig. 4.8 Generated routes for 2 networks with 51 two-way links (102 single-way links)

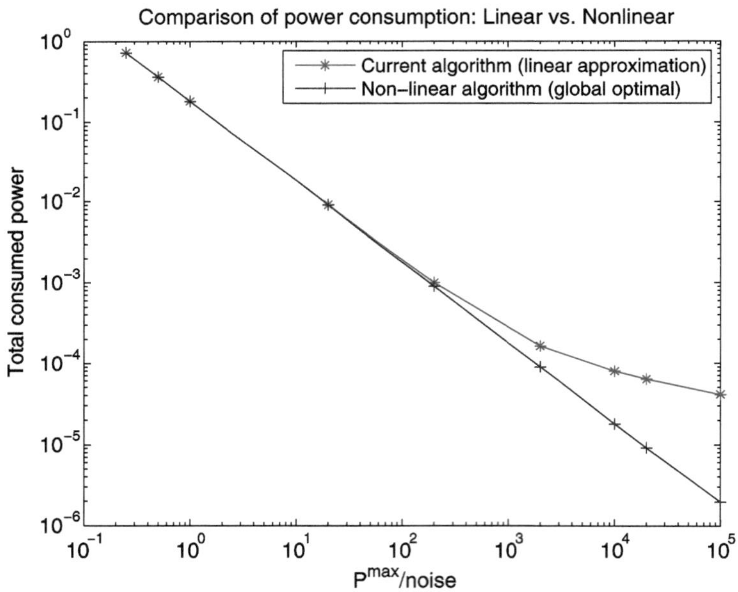

Fig. 4.9 Comparison of the linear approximation with the actual nonlinear model

mon node. In other words, it is assumed that each node has an orthogonal channel for communication, and hence there is no interference between the transmission of two nodes as long as they have separate receiving nodes. However, in our model we assume that interference is a function of distance between the nodes. In this case using separate links in different paths would increase the interference, and hence we may need to consume more power for communication. These results suggest that the interaction between routing, scheduling and power control is very complex and it is very important to consider appropriate physical and MAC layer models in the design of algorithms.

In Fig. 4.9 we ran a series of simulations to test the preciseness of the solution provided by the linear model. We considered a network with one source-destination pair and traffic demand of 200 Kbps. The solution of the nonlinear method is globally optimal, in the sense that it gives the absolute minimum required power to satisfy the traffic demand, i.e. the end-to-end rate from source to destination. The solution of the linear method however solves a power minimization algorithm based on a linear approximation of the capacity formula. Intuitively, we expect that these two solutions be close when the original approximation is tight. The results of our simulations confirm our hypothesis. As can be seen in the plots of Fig. 4.9, the lower the maximum transmitting power, $P^{\max}$ becomes, the closer the two curves will get.

Chapter 5
Connectivity in Wireless Networks

In the previous chapters, we tried to find methods to cope with interference in a wireless network in different layers. However, regardless of what method we use, the interference has impacts on the connectivity of the network. We sure want to have a connected network all the time. The question then will be whether this is possible or not. In this chapter we first define a few connectivity measures and then investigate the connectivity of a network based on those metrics.

1 The Capacity Metric

The discussion of this section revolves around defining a pair of probabilistic and deterministic metrics of connectivity based on the capacity of wireless MIMO channels. While our general probabilistic metric is defined based on an outage capacity analysis for such channels, our ergodic metric provides a simplified deterministic treatment of the probabilistic metric.

Probabilistic Capacity Metric

In this subsection, we introduce our first probabilistic metric of connectivity relying on the concepts of outage capacity and outage probability. Calculating estimates or upper bounds of capacity in the case of uncorrelated and correlated Single-Input Single-Output (SISO) and MIMO channels both with Gaussian and non-Gaussian noise has been the subject of heavy research in the past years. The concept of outage capacity was first introduced in [61]. Outage capacity provides an elegant description of the achievable rate of a communication channel. Simply put, it represents a probabilistic measure of the maximum number of bits per cycle that can be transmitted for a given error rate. The authors of [61] also provided approximations of the capacity of Identically and Independently Distributed (IID) MIMO Rayleigh channels. In [20], methods of calculating the capacity of correlated MIMO channels with

S.J. Kazemitabar, *Coping with Interference in Wireless Networks*, Signals and Communication Technology,
DOI 10.1007/978-90-481-9990-7_5,

Gaussian noise were proposed. The authors of [65] numerically verified that the approximations of capacity derived in [61] work well under various fading conditions in the presence of Rayleigh distributed interference, for a wide Signal-to-Noise Ratio (*SNR*) range, and even when the channel is semi-temporally correlated.

Our discussion below represents a treatment of the subject material relying on the cited literature articles above. In order to be consistent with the literature work of capacity for MIMO channels, the analysis is carried out by explicitly working with the input and output signals of a fading channel.

Consider an ad-hoc topology with q wireless flat fading links $\{\mathcal{L}_1, \ldots, \mathcal{L}_q\}$ on which transmission powers are $\{P_1, \ldots, P_q\}$, respectively. Link i is associated with the i-th transmitter/receiver pair. Each link may be connecting multiple antenna mobile nodes. Suppose, per symbol transmission power P_i is equally distributed among M_i transmit antennas of link i. The number of receive antennas for link i is assumed to be N_i. Further, let us assume that the matrix H_{ij} represents the fading channel between the transmitter of link j and the receiver of link i. Denoting $\mathbf{S}_i$ as the $M_i \times T$ symbol matrix of link i transmitted over T discrete time blocks, the received symbol matrix at link i is the following $N_i \times T$ matrix

$$\mathbf{R}_i = H_{ii}\mathbf{S}_i + \mathbf{\Gamma}_i \tag{5.1}$$

where the channel matrices H_{ii} consist of complex Gaussian random variable elements and $\mathbf{\Gamma}_i = \sum_{j \neq i} H_{ij}\mathbf{S}_j + \mathbf{n}_i$ represents the combined effects of interference and noise. We assume that the receiver of link i knows the channel matrix H_{ii} while the transmitter of link i only knows its distribution. The quantities $(\mathbf{\Gamma}_i | H_{ij})$ can be considered to form a Gaussian random process due to the following lines of reasoning. As discussed in Chap. 2 of [25], we know that the codewords $\mathbf{S}_j$ should be chosen from a Gaussian distribution to be capacity achieving. Further, H_{ij}'s are known at the receiver. Since the elements $H_{ij}\mathbf{S}_j$ are linear combinations of independent Gaussian random variables, they are themselves Gaussian. In addition, any $\mathbf{S}_j$ or $\mathbf{n}_i$ term at any given time slot is independent of its counter parts at other time slots. Since the transmitter does not know the channel, it assigns the codewords independently at each time slot. Therefore, $(\mathbf{\Gamma}_i | H_{ij})$ forms a Gaussian random process. The covariance matrix for the resulting noise term is expressed as

$$\begin{aligned} K_i &= E\{\mathbf{\Gamma}_i . \mathbf{\Gamma}_i{}^{\dagger}\} \\ &= E\left\{\left(\sum_{j \neq i} H_{ij}\mathbf{S}_j + \mathbf{n}_i\right) . \left(\sum_{k \neq i} H_{ik}\mathbf{S}_k + \mathbf{n}_i\right)^{\dagger}\right\} \\ &= E\left\{\sum_{j \neq i} H_{ij}\mathbf{S}_j\mathbf{S}_j^{\dagger}H_{ij}^{\dagger}\right\} + \overline{P}_i^{(n)} I \end{aligned} \tag{5.2}$$

where the superscript $\dagger$ indicates the Hermitian operator, E represents the expectation operator, and $\overline{P}_i^{(n)}$ is the average power of noise. Since we are assuming that

H_{ij} coefficients are known at the receiver,

$$K_i = \sum_{j \neq i} H_{ij} E\{\mathbf{S}_j \mathbf{S}_j^\dagger\} H_{ij}^\dagger + \overline{P}_i^{(n)} I$$
$$= \sum_{j \neq i} H_{ij} \Phi_j H_{ij}^\dagger + \overline{P}_i^{(n)} I \quad (5.3)$$

where I is the identity matrix and Φ_j indicates the covariance matrix of the input signal vector of link j. Then, the mutual information $\mathcal{I}$ between $\mathbf{S}_i$ and $\mathbf{R}_i$ is derived as[1]

$$\mathcal{I}(\mathbf{S}_i; \mathbf{R}_i) = \log_2 \det\big(I + K_i^{-1} H_{ii} \Phi_i H_{ii}^\dagger\big) \quad (5.4)$$

To find the capacity, one needs to maximize $\mathcal{I}(\mathbf{S}_i; \mathbf{R}_i)$ subject to a transmission power constraint $Tr(\Phi_i) \leq P_i$ on link i where $Tr(\Phi_i)$ and P_i denote the trace of Φ_i and the transmission power of link i, respectively.

The choice of covariance matrix achieving the capacity in (5.4) depends on the realization of the channel matrix. When the channel is not known at the transmitter, the best strategy is to distribute the input power equally among the transmit antennas. The latter results in a covariance matrix Φ_i that is a multiple of the identity matrix. Considering the constraint $Tr(\Phi_i) \leq P_i$, we have $\Phi_i = \frac{P_i}{M_i} I$ resulting in the following capacity determination

$$C_i = \log_2 \det\left(I + \frac{P_i}{M_i} K_i^{-1} H_{ii} H_{ii}^\dagger \right) \text{ bps/Hz} \quad (5.5)$$

Note that the capacity can be expressed in terms of a natural logarithm rather than a base 2 logarithm assuming the unit of measurement is changed from bps/Hz to nats/s/Hz.

In the most general case, the capacity expression of (5.5) can be only calculated numerically. When the number of links is relatively large, one can utilize central limit theorem to conclude that the covariance matrix of (5.3) can be expressed as a multiple of the identity matrix. The reasoning follows. Relying on the equation $\Phi_i = \frac{P_i}{M_i} I$, we note that the first term of the covariance matrix K_i is in the form of $\sum_{j \neq i} \frac{P_j}{M_j} H_{ij} H_{ij}^\dagger$. Since the non-diagonal entries of the first term are the sum of zero-mean random variables, central limit theorem implies that they tend to the mean value of the random variables, zero. Further, the diagonal entries of the first term consist of the sum of the square of the magnitudes of the channel coefficients from interfering links. Consequently, they represent the power of interfering signals. Thus, the covariance matrix of (5.3) is expressed in the following form

$$K_i \simeq \big[\overline{P}_i^{(I)} + \overline{P}_i^{(n)}\big] I \quad (5.6)$$

[1]The symbol $\mathcal{I}$ used to denote mutual information should be distinguished from the symbol I to denote the identity matrix.

where $\overline{P}_i^{(I)}$ and $\overline{P}_i^{(n)}$ are the average power of interference and noise, respectively. Therefore, (5.5) can be rewritten as follows

$$C_i \simeq \log_2 \det\left(I + \frac{\overline{SINR}_i}{M_i} H_{ii} H_{ii}^{\dagger}\right) \text{ bps/Hz} \tag{5.7}$$

with $\overline{SINR}_i$ denoting the average signal-to-interference-noise-ratio. Next, we note that the capacity in (5.7) is defined for a fixed realization of the fading channel H_{ii} at link i over a large block length. Every realization of the channel has some probability attached to it through the statistical model of H_{ii}. We assume that the matrix H_{ii} consists of zero-mean Gaussian random variables, i.e., each element of the matrix has a fading envelope described by Rayleigh distribution. It is well known [66] that the sum of q zero-mean IID complex Gaussian random variables with a standard deviation $\frac{1}{\sqrt{2\lambda}}$ is a zero-mean Gaussian random variable with a standard deviation $\sqrt{\frac{q}{2\lambda}}$. Since the channel matrices H_{ii} are random in nature, the capacity in (5.7) can be treated as a random variable.

According to Singular Value Decomposition (SVD) theorem, C_i can be calculated in terms of the positive eigenvalues of $H_{ii} H_{ii}^{\dagger}$ as

$$C_i \simeq \sum_{l=1}^{\rho} \log_2\left[1 + \frac{\overline{SINR}_i}{M_i} \sigma_l\right] \text{ bps/Hz} \tag{5.8}$$

where σ_l's with $l \in \{1, \ldots, \rho\}$ denote the positive eigenvalues of $H_{ii} H_{ii}^{\dagger}$ and ρ is the rank of H_{ii}. Therefore, the capacity C_i represents a scalar function of the set of random variables $\{\sigma_1, \ldots, \sigma_\rho\}$. Our work of [48] describes how the Probability Density Function (PDF) of capacity can be calculated depending on the values of M_i and N_i. Here, we summarize the results. The PDF of $H_{ii} H_{ii}^{\dagger}$ for the case of $M_i \times N_i = 1 \times 1$ is described in the form of

$$f_z(z) = \lambda e^{-\lambda z} \tag{5.9}$$

The PDF identified above represents the only positive eigenvalue of the scalar function $H_{ii} H_{ii}^{\dagger}$. For the cases of $M_i \times N_i = 2 \times 1$ and $M_i \times N_i = 1 \times 2$, the PDF of $H_{ii} H_{ii}^{\dagger}$ is expressed as

$$f_z(z) = \lambda^2 z e^{-\lambda^2 z} \tag{5.10}$$

Again, the PDF identified above represents the only positive eigenvalue of $H_{ii} H_{ii}^{\dagger}$. The results for the case of $M_i \times N_i = 2 \times 2$ are numerically calculated similar to the case of $M_i \times N_i = 2 \times 1$ with an H_{ii} matrix consisting of four pairs of complex Gaussian random variables.

Treating capacity as a random variable with a given PDF provides us with an opportunity to represent a novel connectivity metric based on the concept of outage capacity. We introduce our first metric of connectivity as

$$Pr(C_i < C_{\text{out}}) \leq \Delta_C \tag{5.11}$$

where $Pr(.)$, C_{out}, and Δ_C represent probability, the threshold of connectivity also known as outage capacity, and the outage probability, respectively. While our definition of outage matches that of [61], it differs slightly from that of [20]. According to [20], the outage is defined as

$$\inf_{Tr(\Phi_i)\le P_i} Pr(C_i < C_{\text{out}}) \le \Delta_C \tag{5.12}$$

The main difference between the two definitions is that the latter may assign zero power to some of the transmit antennas while the former utilizes all of the antennas. According to our outage capacity metric matching the former definition, two nodes are connected if the probability of the outage event for the link between them is less than a given value.

Deterministic Capacity Metric

In this subsection, we provide a deterministic treatment of the connectivity metric of the previous subsection assuming the underlying wireless channel is ergodic. The ergodic capacity $\overline{C}_i$ of link i can be expressed as [20]

$$\overline{C}_i = E[C_i] \simeq E\left[\log_2 \det\left(I + \frac{\overline{SINR}_i}{M_i} H_{ii} H_{ii}^{\dagger}\right)\right] \text{ bps/Hz} \tag{5.13}$$

Utilizing SVD theorem and the results of random matrix theory, the following expression can be derived for the ergodic capacity of MIMO channels.

$$\overline{C}_i \simeq u \int_0^{\infty} \log_2\left(1 + \frac{\overline{SINR}_i}{M_i} x\right) f_x(x)\, dx \tag{5.14}$$

In (5.14), $f_x(.)$ represents the PDF of a randomly selected eigenvalue of the Wishart matrix defined as

$$f_x(x) = \frac{1}{u} \sum_{k=0}^{u-1} \frac{k! x^{v-u} e^{-x}}{(k+v-u)!} \left[\Lambda_k^{v-u}(x)\right]^2, \quad x \ge 0 \tag{5.15}$$

with parameters $u = \min(M_i, N_i)$ and $v = \max(M_i, N_i)$. In (5.15), $\Lambda_k^m(x)$ denotes the Laguerre polynomial of order k defined as

$$\Lambda_k^m(x) = \frac{e^x x^{-m}}{k!} \frac{d^k}{dx^k}\{e^{-x} x^{k+m}\} = \sum_{h=0}^{k} (-1)^h \binom{k+m}{k+h} \frac{x^h}{h!} \tag{5.16}$$

where $\binom{k+m}{k+h}$ is the binomial coefficient.

In [73], a simplified expression for the ergodic capacity of (5.14) is derived under average transmit power and equal power allocation constraints as

$$\overline{C}_i \simeq e^{\frac{M_i}{\overline{SINR}_i}} \log_2 e \sum_{k=0}^{u-1} \sum_{l=0}^{k} \sum_{m=0}^{2l} \left[\frac{(-1)^m (2l)!(v-u+m)!}{2^{2k-m} l! m! (v-u+l)!} \right.$$
$$\left. \times \binom{2k-2l}{k-l} \binom{2l+2v-2u}{2l-m} \sum_{n=0}^{v-u+m} \Psi_{n+1}\left(\frac{M_i}{\overline{SINR}_i}\right) \right] \tag{5.17}$$

where $\Psi_n(z)$ is defined as

$$\Psi_n(z) = \int_1^{\infty} e^{-zx} x^{-n} \, dx, \quad n = 0, 1, \ldots \tag{5.18}$$

with $Re(z) > 0$.

Using the ergodic capacity of (5.13) or equivalently (5.17), we can introduce a more simplified connectivity metric in the form of

$$\overline{C}_i \geq C_{\text{out}} \tag{5.19}$$

where C_{out} is the threshold of connectivity.

2 The *SER* Metric

The discussion of this section revolves around providing probabilistic and deterministic measures of connectivity based on symbol error rate. Symbol error rate can in turn be related to the characteristics of the underlying communication system such as *SINR*, modulation, and antenna configuration.

Probabilistic SER Metric

Once again, consider the ad-hoc topology described previously consisting of q flat fading wireless links $\{\mathcal{L}_1, \ldots, \mathcal{L}_q\}$ on which transmission powers are $\{P_1, \ldots, P_q\}$, respectively. Associated with each element $H_{ij}(n,m)$ of channel matrices, we define the fading factors $F_{ij}(n,m) = |H_{ij}(n,m)|^2$. For each link i, the *SER* can be derived as an exact function of the average *SINR* and the corresponding fading factors $F_{ii}(n,m)$.

We start by investigating the expressions of *SER* for $1 \times N_i$ link i. In [72], the expressions of *SER* for $1 \times N_i$ link i in terms of the number of signal points in the constellation and the average *SNR* can be found. The calculations of [72] are carried out under the assumption of facing a complex Gaussian noise signal. As one of the operating scenarios, the calculations are carried out for a Rayleigh fading channel and utilizing PSK modulation. Because the quantity of interest is *SINR* rather than *SNR* in the context of current discussion, we need to investigate the effects of interference signals in (5.1). First, we claim that the product $H_{ij}\mathbf{S}_j$ remains Gaussian in (5.1). Using a limited constellation set like BPSK with a uniform distribution for each signal instead of capacity achieving codebook, we verify the claimed statement for the case of 1×1 link. Generalization to L-PSK modulation and $M_i \times N_i$ link is then straightforward. If we use BPSK modulation, $H_{ij}\mathbf{S}_j$ will be a scalar random variable with the following description

$$\mathbf{X}_j = H_{ij}\mathbf{S}_j = \begin{cases} H_{ij}, & \text{with probability } 0.5 \\ -H_{ij}, & \text{with probability } 0.5 \end{cases} \tag{5.20}$$

To see why $\mathbf{X}_j$ is a complex Gaussian random variable, we need to show that both real and imaginary parts of this random variable are normally and independently distributed. Since

$$Re\{\mathbf{X}_j\} = Re\{H_{ij}\mathbf{S}_j\} = \begin{cases} Re\{H_{ij}\}, & \text{with probability } 0.5 \\ -Re\{H_{ij}\}, & \text{with probability } 0.5 \end{cases} \tag{5.21}$$

We have

$$\begin{aligned} Pr(Re\{\mathbf{X}_j\} < x) &= \frac{1}{2}Pr(Re\{H_{ij}\} < x) + \frac{1}{2}Pr(-Re\{H_{ij}\} < x) \\ &= Pr(Re\{H_{ij}\} < x) \end{aligned} \tag{5.22}$$

Therefore, the distribution of the real part of $\mathbf{X}_j$ is normal. Relying on the same argument, the distribution of the imaginary part of $\mathbf{X}_j$ is normal. Next, we observe

$$\begin{aligned} &Pr(Re\{X_j\} < x \mid Im\{X_j\} < y) \\ &\quad = Pr(Re\{X_j\} < x \mid Im\{X_j\} < y, S_j = 1)Pr(S_j = 1) \\ &\qquad + Pr(S_j.Re\{X_j\} < x \mid Im\{X_j\} < y, S_j = -1)Pr(S_j = -1) \\ &\quad = \frac{1}{2}Pr(Re\{X_j\} < x \mid Im\{X_j\} < y, S_j = -1) \\ &\qquad + \frac{1}{2}Pr(-Re\{X_j\} < x \mid -Im\{X_j\} < y, S_j = -1) \\ &\quad = \frac{1}{2}Pr(Re\{X_j\} < x) + \frac{1}{2}P(-Re\{X_j\} < x) \\ &\quad = Pr(Re\{X_j\} < x) \end{aligned} \tag{5.23}$$

which concludes our reasoning of independence.

Since the product $H_{ij}\mathbf{S}_j$ remains Gaussian in (5.1) and the sum of Gaussian random variables is still Gaussian [66], the signal $\mathbf{\Gamma}_i$ can still be treated as Gaussian. However, the resulting Gaussian noise is now colored rather than being white. We note that applying Maximum Likelihood (ML) decoding as utilized by [72] to a colored Gaussian noise results in sub-optimality, i.e., identifying upper bounds of the *SER*. Based on the argument above, the analysis of [72] can still be applied to the case of *SINR* utilizing the model of (5.1) the same way it is applied to *SNR*.

According to Sect. 9.2 of [72], the expressions of *SER* are calculated for a $1 \times N_i$ link i utilizing MRC and BPSK modulation as

$$SER_i \simeq Q\left(\sqrt{2\sum_{n=1}^{N_i} F_{ii}(n,1)\overline{SINR}_i}\right) \tag{5.24}$$

where $\overline{SINR}_i$ is the average received signal-to-interference-noise ratio and the Gaussian Q function is defined as

$$Q(x) = \frac{1}{\sqrt{2\pi}}\int_x^{\infty} \exp\left(-\frac{z^2}{2}\right) dz = \frac{1}{\pi}\int_0^{\pi/2} \exp\left(-\frac{x^2}{2\sin^2\phi}\right) d\phi \tag{5.25}$$

Relying on the discussion of Sect. 4.9 of [25], we note that the MRC expressions of (5.24) can also be applied to Alamouti STBCs of [11] with proper scaling factors. Particularly, the results of a 1×2 link utilizing MRC codes can be applied to the case of a 2×1 link utilizing Alamouti codes. Similarly, the results of a 1×4 link utilizing MRC codes can be applied to the case of a 2×2 link utilizing Alamouti codes.

Utilizing BPSK modulation and under the assumption of facing a Rayleigh fading channel, the symbol error rate of link i can be derived from the latter analysis as

$$SER_i \simeq Q\left(\sqrt{2\eta\Upsilon_i\overline{SINR}_i}\right) \tag{5.26}$$

where $\overline{SINR}_i$ is the average signal-to-interference-noise-ratio of link i and η is a constant that depends on the antenna configuration. While the value of η is 1 for 1×1 and 1×2 links utilizing MRC, it changes to 0.5 for 2×1 and 2×2 links utilizing STBCs of [10]. Further, Υ_i is defined as

$$\Upsilon_i = \sum_{m=1}^{M_i}\sum_{n=1}^{N_i} F_{ii}(n,m) \tag{5.27}$$

Since the quantity SER_i as specified by (5.26) represents a function of random variables, it suffices to examine the distribution of $F_{ii}(n,m)$ in order to obtain fading statistics of SER_i. We start from the case of a single transmit and single receive antenna link, i.e., $F_{ii} = F_{ii}(1,1)$ and $M_i = N_i = 1$. In a 1×1 link case and when the channel matrix is identified by a complex Gaussian noise element, one can conclude that $r_i = |H_{ii}|$ has a marginal Rayleigh density function [66] in the form of

$$p_r(r_i) = \frac{r_i e^{-r_i^2/2\mu_i^2}}{\mu_i^2}, \quad r_i \geq 0 \tag{5.28}$$

where μ_i^2 equals to half of the average power of all of the multipath components. The PDF of F_{ii} can be expressed [66] as

$$p_F(F_{ii}) = \frac{1}{2\sqrt{F_{ii}}} p_r(\sqrt{F_{ii}}) \tag{5.29}$$

Once the PDFs of F_{ii} terms are calculated and assuming they are spatially uncorrelated, the PDF of Υ_i is specified [66] as defined in (5.27). Finally, the PDF of SER_i as defined in (5.26) can be numerically calculated in terms of the PDFs of random variables Υ_i.

Having captured the distribution of the symbol error rate for a MIMO link, we are now ready to express our second metric of connectivity in terms of the quantities of interest. We introduce our second metric of connectivity as

$$Pr(SER_i > S_{\text{out}}) \leq \Delta_S \tag{5.30}$$

where $Pr(.)$, S_{out}, and Δ_S represent probability, the threshold of connectivity, and the outage probability, respectively.

Deterministic SER Metric

In this subsection, we assume that the time-varying fading wireless channel is ergodic. Under the assumption of facing an ergodic Rayleigh channel and utilizing BPSK modulation, the random variable SER_i of (5.26) can be substituted by its average value $\overline{SER_i}$ defined as

$$\overline{SER_i} \simeq \int_0^{\infty} \frac{1}{\pi} \int_0^{\pi/2} \exp\left(-\frac{2\eta \Upsilon_i \overline{SINR_i}}{2\sin^2 \tau}\right) d\tau \, p_\Upsilon(\Upsilon_i)\, d\Upsilon_i \tag{5.31}$$

where $p_\Upsilon(.)$ is the PDF of the random variable Υ_i. The result is also valid for L-PSK modulation as

$$\overline{SER_i} \simeq \int_0^{\infty} \frac{1}{\pi} \int_0^{\frac{(L-1)\pi}{L}} \exp\left(-\frac{2\eta \Upsilon_i \overline{SINR_i}}{2\sin^2 \tau}\right) d\tau \, p_\Upsilon(\Upsilon_i)\, d\Upsilon_i \tag{5.32}$$

In [75], closed-form expressions of the integral of (5.32) are calculated. The expressions describe the symbol error rate of a MIMO channel in terms of the number of signal points in the constellation and the average signal-to-noise ratio. We carry out our calculations under the assumption of facing a slow fading ergodic Rayleigh channel and utilizing the PSK modulation scheme. In what follows, we provide the results of our calculations considering the fact that in the current discussion the quantity of interest is $\overline{SINR}$ rather than $\overline{SNR}$. First, we introduce the symbol error rate of a $1 \times N_i$ link i using MRC as

$$\overline{SER_i} \simeq \frac{L_i - 1}{L_i} - \frac{1}{\pi}\sqrt{\frac{\vartheta_i}{1+\vartheta_i}}$$

$$\times \left\{ \left(\frac{\pi}{2} + \tan^{-1}\theta_i \right) \sum_{j=0}^{N_i-1} \binom{2j}{j} \frac{1}{[4(1+\vartheta_i)]^j} \right.$$
$$+ \sin(\tan^{-1}\theta_i) \sum_{j=1}^{N_i-1} \sum_{k=1}^{j} \frac{\zeta_{kj}}{(1+\vartheta_i)^j}$$
$$\left. \times [\cos(\tan^{-1}\theta_i)]^{2(j-k)+1} \right\} \tag{5.33}$$

where $\vartheta_i = \overline{SINR}_i \sin^2(\frac{\pi}{L_i})$, $\theta_i = \sqrt{\frac{\vartheta_i}{1+\vartheta_i}} \cot \frac{\pi}{L_i}$ and $\zeta_{kj} = \frac{\binom{2j}{j}}{\binom{2(j-k)}{j-k} 4^i [2(j-k)+1]}$.

Noting that the number of bits per symbol is related to the number of signal points in the constellation L_i as $\log_2 L_i$, the result of (5.33) for a 1×1 link utilizing BPSK modulation with $L_i = 2$ is expressed as

$$\overline{SER}_i \simeq \frac{1}{2}\left(1 - \sqrt{\frac{\overline{SINR}_i}{1+\overline{SINR}_i}}\right) \tag{5.34}$$

Similarly, the result of (5.33) for a 1×2 link utilizing BPSK modulation is expressed as

$$\overline{SER}_i \simeq \frac{1}{2}\left[1 - \sqrt{\frac{\overline{SINR}_i}{1+\overline{SINR}_i}}\left(1 + \frac{1}{2(1+\overline{SINR}_i)}\right)\right] \tag{5.35}$$

We observe that the symbol error rate of a 1×2 link is improved compared to that of a 1×1 link due to the receive diversity gain.

Further, we note that the symbol error rate of Alamouti STBCs of [10] and [11] can be derived from Sect. 2.1 of [75]. Based on that discussion and for a fixed transmit power, one can derive the symbol error rate of a 2×1 link by replacing $\overline{SINR}_i$ with $\frac{\overline{SINR}_i}{2}$ in (5.35). Similarly, one can obtain the symbol error rate of a 2×2 link by using the results of a 1×4 link. With the choice of BPSK modulation, the result for a 2×2 link is expressed as

$$\overline{SER}_i \simeq \frac{1}{2} - \frac{1}{2}\sqrt{\frac{\overline{SINR}_i}{2+\overline{SINR}_i}}\left(\sum_{j=0}^{3} \binom{2j}{j} \frac{1}{[2(2+\overline{SINR}_i)]^j}\right) \tag{5.36}$$

Using the ergodic symbol error rates above, we can introduce a simplified connectivity metric in the form of

$$\overline{SER}_i \leq S_{\text{out}} \tag{5.37}$$

where S_{out} is the threshold of connectivity.

3 Capturing Temporal Correlation of Ergodic Channels

Up until now, we have assumed that the wireless fading channel is flat implying no temporal correlation exists among the symbols of a single frame. In this section, we capture the effects of temporal correlation on our connectivity metrics. We propose a scheme in which the temporally correlated fading channel is modeled by a finite-state Markov chain. Our scheme can be applied to the cases of our ergodic connectivity metrics. Capturing temporal correlation in the cases of our probabilistic metrics is more complex and the subject of our future study.

A finite-state Markov chain is a discrete-time representation of the behavior of a random variable. Each state is associated with an average quantity representing the value of the random variable at that state. The chain is fully specified by the set of average per state quantities and a set of per state steady-state probabilities. For an S-state chain, S per state average quantities and S steady-state probabilities can be calculated by partitioning the PDF of a random variable with a set of threshold values $\{\xi_0, \ldots, \xi_S\}$ associated with the observed behavior of the random variable.

While such Markov chain modeling approach can be applied to any number of states, we note that utilizing a larger number of states improves the accuracy of the model at the cost of increasing the complexity of calculations. We propose the use of a two-state chain with two states G and B to address the tradeoff between computational complexity and model accuracy. Utilizing a two-state Markov chain applied to the PDF of C_i in (5.8), the ergodic connectivity metric of (5.19) still holds when the ergodic average $\overline{C}_i$ of link i is expressed as

$$\overline{C}_i = \pi_{i,G}^{(C)} \overline{C}_{i,G} + \pi_{i,B}^{(C)} \overline{C}_{i,B} \tag{5.38}$$

where $\pi_{i,G}^{(C)}$ and $\pi_{i,B}^{(C)}$ represent per state steady-state probabilities derived from the ratios of surface integrals of the PDF of C_i within appropriate threshold bounds. Further, average per state quantities $\overline{C}_{i,G}$ and $\overline{C}_{i,B}$ are specified as the ratios of expectation integrals and surface integrals of the PDF of C_i within appropriate threshold bounds [76]. Similarly and utilizing a two-state Markov chain applied to the PDF of SER_i in (5.26), the ergodic connectivity metric of (5.37) still holds when the ergodic average $\overline{SER}_i$ of link i is expressed as

$$\overline{SER}_i = \pi_{i,G}^{(SER)} \overline{SER}_{i,G} + \pi_{i,B}^{(SER)} \overline{SER}_{i,B} \tag{5.39}$$

with similar description of quantities in (5.38) derived from the PDF of SER_i in (5.26) instead of those derived from the PDF of C_i in (5.8).

4 Experimental Verification of Analysis

This section is dedicated to the evaluation of effectiveness for the information theoretic metrics of connectivity introduced in previous sections. It consists of two subsections. The objective of the first subsection is to show why utilizing a metric purely based on *SINR* cannot capture the reality of connectivity in fading ad-hoc networks. The objective of the second subsection is to investigate the results of applying the metrics of previous sections to a random network topology.

Justification of Proposed Connectivity Metrics

We open this section by providing a justification of using our proposed metrics of connectivity instead of the *SINR* metric.

For different choices of antenna configurations of a given link, Fig. 5.1 depicts normalized values of $1 - CDF(SER)$ versus S_{out} where $CDF(SER)$ indicates the Cumulative Distribution Function (CDF) of *SER*. For a choice of S_{out} on the horizontal axis of each figure, the corresponding value on the vertical axis represents the value $Pr(SER > S_{\text{out}})$. Thus, the connectivity metric may be satisfied if the horizontal line representing Δ_S is located above the value of $Pr(SER > S_{\text{out}})$. The figure reveals the fact that the probabilistic measure of the symbol error rate can be different for the same threshold S_{out} based on the antenna configuration. For example, for the choice of $(S_{\text{out}}, \Delta_S) = (0.15, 0.1)$, the connectivity metric of (5.30) is satisfied for 2×1, 1×2, and 2×2 antenna configurations but not 1×1 antenna configuration. While not shown here, similar results are observed in the case of outage capacity metric of connectivity.

For an isolated point-to-point transmission scenario and different antenna configurations, Fig. 5.2 depicts $\overline{C}$ as defined in (5.17) versus $\overline{SINR}$. We note that in an isolated point-to-point communication scenario, there is no interference term and as a result the term $\overline{SINR}$ is the same as $\overline{SNR}$. The figure reveals that the capacity can be different for the same signal strength based on the antenna configuration. For example for the choice of $\overline{SINR} = 10$ dB and $C_{\text{out}} = 5$ bps/Hz in Fig. 5.2, the connectivity metric of (5.19) is only satisfied for 2×2 wireless links but not the other antenna configurations.

Hence, depending on the thresholds of connectivity C_{out}, S_{out}, Δ_C, and Δ_S that are determined by the computing platform of a mobile node, a pure measurement of the signal strength such as *SINR* is not sufficient to capture the connectivity phenomenon.

Connectivity Experiments

In this subsection, we apply our proposed connectivity metrics to a moderate size random ad-hoc topology. In order to provide a meaningful basis of comparison, we compare our results for the same random topology. In our random topology, 200 nodes are distributed on a 2-D domain with an area of 1000 square meters according to a Poisson point process [64]. When measuring connectivity, we assume all of the nodes can transmit at the same time. We note that the use of Gaussian

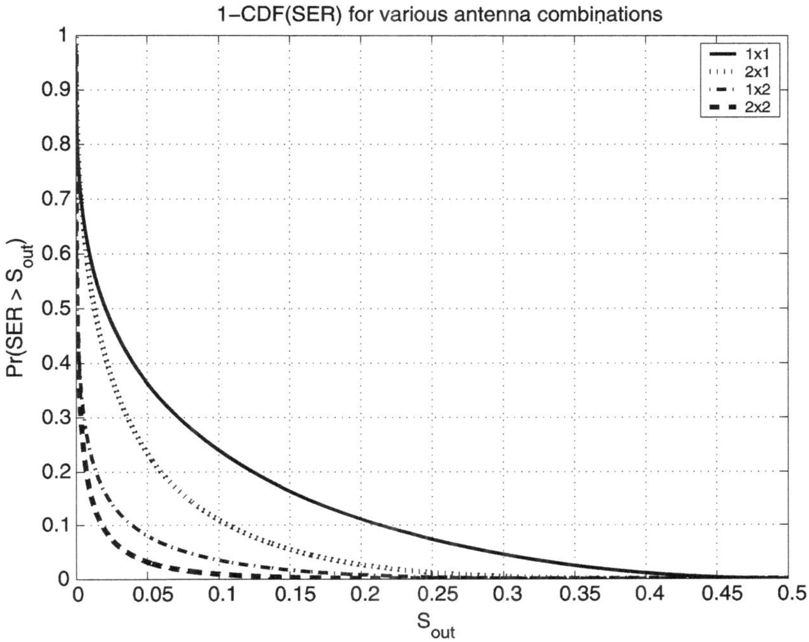

Fig. 5.1 Normalized BPSK plots of $1 - CDF(SER)$ versus S_{out} for a wireless link utilizing different antenna configurations with $SINR = 3$ dB

approximation according to central limit theorem is justified for different antenna configurations as the result of allowing simultaneous transmissions and considering the number of interfering nodes. More specifically, we have conducted a number of experiments to quantify the number of independent interference terms necessary for accurate use of Gaussian approximation. Our experiments have revealed that the use of Gaussian approximation is acceptable with an accuracy of 0.01% when the number of interference terms, each term contributed by an individual path, exceeds 30. The latter translates to 30 nodes in the case of 1×1 antenna configurations and no more than 8 nodes in the case of 2×2 antenna configurations.

The following describes general settings of our experiments. All of the nodes are assumed to be utilizing BPSK modulation. In our probabilistic experiments, we assume that the slow fading wireless channel characterized by a Rayleigh distribution is quasi-static and flat implying there is no temporal correlation between a pair of symbols belonging to the same frame. In our ergodic experiments, we utilize a two-state Markov chain when partitioning the PDFs of the random variables associated with capacity and symbol error rate. In both cases, we set the partitioning thresholds as $\{\xi_0, \xi_1, \xi_2\} = \{0, 1.2039, 10\}$. We assume that each node utilizes a total transmission power of $P = 1W$ on the combined set of its outgoing links. In the case of multiple antenna nodes, the total transmission power is split equally among the antenna paths, i.e., M_i signals are transmitted simultaneously from the M_i transmit antennas at each time slot using Alamouti STBCs of [10] and [11]. The

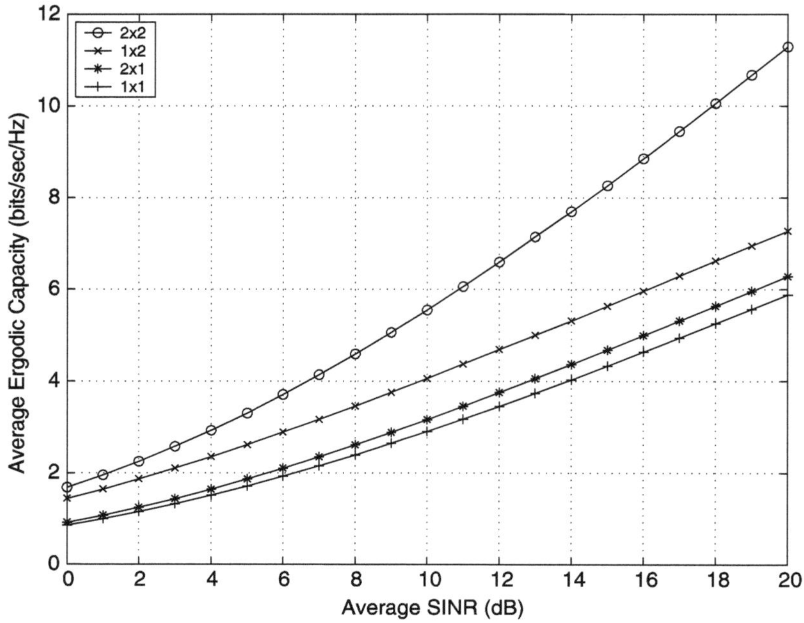

Fig. 5.2 BPSK plots of $\overline{C}$ versus $\overline{SINR}$ for an isolated wireless link utilizing different antenna configurations

expected value of the noise power on each path is assumed to be 10 μW. Depending on a specific experiment, a pair of nodes are considered to form a direct link if one of the probabilistic connectivity metrics of (5.11) and (5.30) or one of the ergodic connectivity metrics of (5.19) and (5.37) holds. A direct link is formed only if both of its nodes can transmit and receive from each other under a connectivity criterion.

For the random topology described above, we consider three scenarios. In the first scenario which serves as our base line SISO scenario, the network is only accommodating single antenna mobile nodes. We refer to this scenario as the 1×1 case. In the second scenario exactly half of the mobile nodes are randomly selected to be equipped with double antennas. We refer to this scenario as the HYBRID case. In the third scenario to which we refer as the 2×2 case, the network is only accommodating double antenna mobile nodes.

We provide the results of our experiments in the case of the probabilistic measures of (5.11) and (5.30) as well as ergodic measures of (5.19) and (5.37). It is important to note that all of our measures implicitly capture the effects of shadowing and distance in addition to fading. The latter is due to the fact that the measures are all expressed as a function of average *SINR*. We refer the reader to the work of [76] to see how shadowing, distance, and fading are captured in the calculations of average *SINR*. For the random topology network above, Fig. 5.3 includes sample connectivity graphs chosen from among a large set of experiments run with different combinations of simulation parameters. While the parameter settings of the graphs

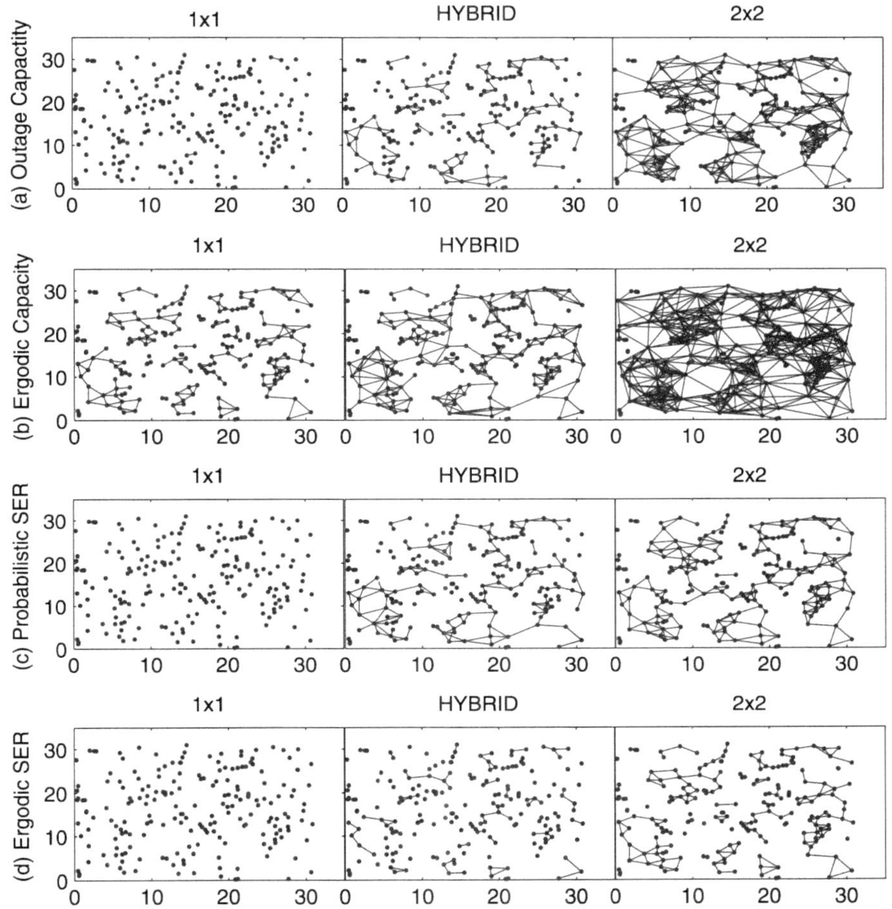

Fig. 5.3 Connectivity graphs of a random topology network in a square domain of 1000 square meters. The *columns from left to right* correspond to single antenna, hybrid, and double antenna mobile nodes. (**a**) The *illustrations of the first row* show the results of utilizing probabilistic connectivity metric of (5.11) with $C_{\text{out}} = 2$ bps/Hz and $\Delta_C = 0.01$. (**b**) The *illustrations of the second row* show the results of utilizing ergodic connectivity metric of (5.19) with $C_{\text{out}} = 4$ bps/Hz. (**c**) The *illustrations of the third row* show the results of utilizing probabilistic connectivity metric of (5.30) with $S_{\text{out}} = 0.02$ and $\Delta_S = 0.01$. (**d**) The *illustrations of the fourth row* show the results of utilizing ergodic connectivity metric of (5.37) with $S_{\text{out}} = 0.0001$

merely represent our sampling choices, the results of all of our experiments remain consistent. We refer the reader to our related work of [47] where we investigate the effects of parameter variations on connectivity.

Reviewing the connectivity graphs, we observe that they vary depending on not only the *SINR* measure but modulation, antenna configurations, and other settings of the nodes. While a pure measurement of the signal strength such as *SINR* is not quite capable of describing the phenomenon of connectivity, utilizing one of our proposed

Table 5.1 A comparison of the relative sizes of the largest connected cluster utilizing outage capacity connectivity metric

	$C_{out} = 2$ $\Delta_C = 0.01$	$C_{out} = 1.5$ $\Delta_C = 0.01$	$C_{out} = 1$ $\Delta_C = 0.02$
1×1	1.5%	2%	7%
HYBRID	12.0%	31.0%	90.5%
2×2	90.5%	94.0%	98.0%

Table 5.2 A comparison of the relative sizes of the largest connected cluster utilizing ergodic capacity connectivity metric

	$C_{out} = 3.8$	$C_{out} = 4$
1×1	32.5%	17.5%
HYBRID	85.5%	85.5%
2×2	98.0%	98.0%

Table 5.3 A comparison of the relative sizes of the largest connected cluster utilizing probabilistic *SER* connectivity metric

	$S_{out} = 0.01$ $\Delta_S = 0.01$	$S_{out} = 0.02$ $\Delta_S = 0.01$	$S_{out} = 0.05$ $\Delta_S = 0.02$
1×1	1.5%	1.5%	7%
HYBRID	29.5%	33.0%	87.5%
2×2	68.5%	83.5%	92.5%

Table 5.4 A comparison of the relative sizes of the largest connected cluster utilizing ergodic *SER* connectivity metric

	$S_{out} = 0.0001$	$S_{out} = 0.01$
1×1	1%	8.0%
HYBRID	4.0%	87.0%
2×2	17.5%	92.5%

metrics provides a better way of properly capturing the effects of the quantities of interest when investigating connectivity.

From the results of the experiments, we can also calculate the percentages of the nodes belonging to the largest connected cluster of nodes. The larger the percentage, the closer the network to being fully connected and a measure of 100% is associated with a fully connected network. Utilizing the connectivity metrics of (5.11) and (5.19), Tables 5.1 and 5.2 report the connectivity results for different combination of choices of C_{out} and Δ_C with similar other settings. We observe that decreasing the value of C_{out} and increasing the value of Δ_C increases the size of the largest cluster of the connectivity graph.

Similarly, utilizing the *SER* connectivity metrics of (5.30) and (5.37), Tables 5.3 and 5.4 report the results for different combination of choices of S_{out} and Δ_S with similar other settings. We notice that increasing the values of S_{out} and Δ_S increases the size of the largest cluster observed in the connectivity graph.

At the end of this section, it is worth mentioning that the calculation costs of our measures of connectivity, accrued at each node and particularly under mobility, are relatively higher than those of the distance and *SINR* measures. Nonetheless, we believe that the increased computational cost of our approach is justified in order to create accurate benchmarks capable of truly capturing the connectivity phenomenon. While this work has proposed analytical measures for studying connectivity, we are currently working on developing intelligent schemes resulting in the reduction of the calculation costs of our measures under mobility. Finally, we would like to point out that the existence of error recovery and scheduling schemes utilized by Medium Access Control (MAC) technologies at the link layer can and will affect connectivity. Capturing the combined effects of PHY and MAC layers is outside the scope of this study and is the subject of future studies.

References

1. V. Tarokh, A. Naguib, N. Seshadri, A.R. Calderbank, Combined array processing and space-time coding. IEEE Trans. Inf. Theory **45**, 1121–1128 (1999)
2. T.M. Cover, Broadcast channels. IEEE Trans. Inf. Theory **18**, 2–14 (1972)
3. T.M. Cover, An achievable rate region for the broadcast channel. IEEE Trans. Inf. Theory **21**, 399–404 (1975)
4. T.M. Cover, A. El Gamal, Capacity theorems for the relay channel. IEEE Trans. Inf. Theory **25**, 572–584 (1979)
5. L. Li, A. Goldsmith, Capacity and optimal resource allocation for fading broadcast channels: Part I ergodic capacity. IEEE Trans. Inf. Theory **47**, 1083–1102 (2001)
6. L. Li, A. Goldsmith, Capacity and optimal resource allocation for fading broadcast channels: Part II outage capacity. IEEE Trans. Inf. Theory **47**, 1103–1127 (2001)
7. D. Tse, S.V. Hanly, Multiaccess fading channels part I: polymatroid structure, optimal resource allocation and throughput capacities. IEEE Trans. Inf. Theory **44**(7), 2796–2815 (1998)
8. D. Tse, S.V. Hanly, Multiaccess fading channels part II: delay-limited capacities. IEEE Trans. Inf. Theory **44**(7), 2816–2831 (1998)
9. V. Tarokh, N. Seshadri, A.R. Calderbank, Space-time codes for high data rate wireless communication: performance criterion and code construction. IEEE Trans. Inf. Theory **44**(2), 744–765 (1998)
10. S.M. Alamouti, A simple transmitter diversity scheme for wireless communications. IEEE J. Select. Areas Commun. **16**, 1451–1458 (1998)
11. V. Tarokh, H. Jafarkhani, A.R. Calderbank, Space-time block codes from orthogonal designs. IEEE Trans. Inf. Theory **45**, 1456–1467 (1999)
12. L. Zheng, D. Tse, Diversity and multiplexing: a fundamental tradeoff in multiple-antenna channels. IEEE Trans. Inf. Theory **49**(5), 1073–1096 (2003)
13. D. Tse, P. Viswanath, L. Zheng, Diversity-multiplexing tradeoff in multiple access channels. IEEE Trans. Inf. Theory **50**(9), 1859–1874 (2004)
14. K. Azarian, H. El Gamal, P. Schinter, Achievable diversity-vs-multiplexing tradeoffs in half-duplex cooperative channels, in *ITW 2004* (2004), pp. 292–297
15. A.F. Naguib, N. Seshadri, A.R. Calderbank, Applications of space-time block codes and interference suppression for high capacity and high data rate wireless systems, in *Proc. 32nd Asilomar Conf. Signals, Systems and Computers* (1998), pp. 1803–1810
16. N. Al-Dhahir, A.R. Calderbank, Further results on interference cancellation and space-time block codes, in *Proc. 35th Asilomar Conf. on Signals, Systems and Computers* (2001), pp. 257–262
17. H. Jafarkhani, A quasi-orthogonal space-time block code. IEEE Trans. Commun. **49**(1), 1–4 (2001)

S.J. Kazemitabar, *Coping with Interference in Wireless Networks*, Signals and Communication Technology,
DOI 10.1007/978-90-481-9990-7, © Springer Science+Business Media B.V. 2011

18. W. Su, X. Xia, Quasi-orthogonal space-time block codes with full diversity. IEEE Trans. Inf. Theory **50**(10), 2331–2347 (2004)
19. P.W. Wolniansky, G.J. Foschini, G.D. Golden, R.A. Valenzuela, V-BLAST: an architecture for realizing very high data rates over the rich-scattering wireless channel, in *Proc. ISSSE-98*, Pisa, Italy, September 29, 1998. Invited paper
20. E. Telatar, Capacity of multiantenna Gaussian channels. AT&T-Bell Lab. Internal Tech. Memo (1995)
21. G.J. Foschini, M.J. Gans, On limits of wireless communication in a fading environment when using multiple antennas. Bell Lab. Tech. J. **1**(2), 204–210 (1996)
22. J.-C. Belfiore, G. Rekaya, E. Viterbo, The golden code: a 2×2 full-rate space-time code with non-vanishing determinants. IEEE Trans. Inf. Theory **51**(4), 310 (2005)
23. F.E. Oggier, G. Rekaya, J.-C. Belfiore, E. Viterbo, Perfect space-time block codes. IEEE Trans. Inf. Theory **52**(9), 3885–3902 (2006)
24. X. Liang, X. Xia, On the non-existence of rate-one generalized complex orthogonal designs. IEEE Trans. Inf. Theory **49**(11), 2984–2988 (2003)
25. H. Jafarkhani, *Space-Time Coding: Theory and Practice* (Cambridge University Press, Cambridge, 2005)
26. Y. Zhu, H. Jafarkhani, Differential modulation based on quasi-orthogonal codes. IEEE Trans. Wirel. Commun. **4**(6), 3018–3030 (2005)
27. C. Yuen, Y.L. Guan, T.T. Tjhung, Quasi-orthogonal STBC with minimum decoding complexity. IEEE Trans. Wirel. Commun. **4**(5) (2005)
28. J. Kazemitabar, H. Jafarkhani, Multiuser interference cancellation and detection for users with four transmit antennas, in *Proc. of ISIT06* (2006), pp. 1914–1918
29. J. Kazemitabar, H. Jafarkhani, Multiuser interference cancellation and detection for users with more than two transmit antennas. IEEE Trans. Commun. **56**(4), 574–583 (2008)
30. S. Sirianunpiboon, S.D. Howard, A.R. Calderbank, Diversity gains across line of sight and rich scattering environments from space-polarization-time codes, in *Proc. ITWN07* (2007), pp. 1–5
31. D. Tse, P. Viswanath, *Fundamentals of Wireless Communication* (Cambridge University Press, Cambridge, 2005)
32. H. Blocskei, A.J. Paulraj, Performance of space-time codes in the presence of spatial fading correlation, in *Asilomar Conf. Signals, Syst. Comput. 2000*, 29 October–1 November 2000, pp. 687–693
33. J.R. Silvester, Determinants of block matrices. http://www.mth.kcl.ac.uk/~jrs/gazette/blocks.pdf. Accessed 10 Oct. 2010
34. Degen's Eight-Square Identity. http://mathworld.wolfram.com/DegnsEight-SquareIdentity.html. Accessed 10 Oct. 2010
35. N. Shawash, Derivation of matrix inversion lemma. http://www.stanford.edu/~wonghoi/ref/Inverse%5B1%5D.pdf. Accessed 10 Oct. 2010
36. R. Cruz, A. Sanathanam, Optimal routing, link scheduling and power control in multi-hop wireless networks, in *Proc. IEEE INFOCOM*, 2003
37. A. Sanathanam, Joint optimization of radio resources in wireless multi-hop networks. Ph.D. thesis, University of California, San Diego
38. M. Kodialam, T. Nandagopal, Characterizing achievable rates in multi-hop wireless mesh networks with orthogonal channels. IEEE/ACM Trans. Netw. **13**(4), 868–880 (2005)
39. R. Bahita, M. Kodialam, On power efficient communication over multi-hop wireless networks: joint routing, scheduling and power control, in *Proc. IEEE INFOCOM*, 2004
40. L. Lin, X. Lin, N. Shroff, Low-complexity and distributed energy minimization in multi-hop wireless networks, in *Proc. INFOCOM 2007*, 2007
41. M.J. Neely, Energy optimal control for time varying wireless networks. IEEE Trans. Inf. Theory **52**(2), 2915–2934 (2006)
42. V. Bharghavan, A. Demers, S. Shenker, L. Zhang, MACAW: a media access protocol for wireless LAN's, in *Proc. ACM SIGCOMM'94*, 1994
43. M. Garetto, J. Shi, E.W. Knightly, Modeling media access in embedded two-flow topologies of multi-hop wireless networks, in *Proc. Mobicom'05*, 2005

44. X. Lin, N. Shroff, The impact of imperfect scheduling on cross-layer rate control in multihop wireless networks, in *Proc. IEEE Infocom'05*, 2005
45. J. Kazemitabar, H. Yousefi'zadeh, H. Jafarkhani, Impact of physical layer parameters on connectivity of ad hoc networks, in *Proc. IEEE ICC*, 2006
46. H. Jafarkhani, H. Yousefi'zadeh, J. Kazemitabar, Capacity-based connectivity of MIMO fading ad hoc networks, in *Proc. IEEE Globecom*, 2005
47. H. Yousefi'zadeh, H. Jafarkhani, J. Kazemitabar, Outage probability metrics of connectivity for MIMO fading ad-hoc networks, in *Proc. IEEE Milcom*, 2005
48. H. Yousefi'zadeh, H. Jafarkhani, J. Kazemitabar, A study of connectivity in MIMO fading ad-hoc networks. J. Commun. Netw. **11**(1), 47–56 (2009)
49. J.E. Kelly Jr., The cutting plane method for solving convex programs. J. Soc. Ind. Appl. Math. **8**(4), 418–429 (1960)
50. http://www.gams.com. Accessed 10 Oct. 2010
51. http://archimedes.cheme.cmu.edu/baron/baron.html. Accessed 10 Oct. 2010
52. S. Boyd, L. Vanderberghe, *Convex Optimization* (Cambridge University Press, Cambridge, 2004)
53. F. Baccelli, B. Blaszczyszyn, On a coverage process ranging from the Boolean model to the Poisson Voronoi tessellation, with applications to wireless communications. Adv. Appl. Probab. **33**(2), 293–323 (2001)
54. C. Bettstetter, On the minimum node degree and connectivity of a wireless multihop network, in *Proc. ACM MOBIHOC*, 2002
55. L. Booth, J. Bruck, M. Franschetti, R. Meester, Continuum percolation and the geometry of wireless networks. Ann. Appl. Probab. **13**(2), 722–731 (2003)
56. Y.-C. Cheng, T.G. Robertazzi, Critical connectivity phenomena in multihop radio models. IEEE Trans. Commun. **37**(7), 770–777 (1989)
57. O. Dousse, P. Thiran, M. Hasler, Connectivity in ad-hoc and hybrid networks, in *Proc. IEEE INFOCOM*, 2002
58. O. Dousse, F. Baccelli, P. Thiran, Impact of interferences on connectivity in ad-hoc and networks, in *Proc. IEEE INFOCOM*, 2003
59. G.J. Foschini, M.J. Gans, On limits of wireless communication in a fading environment when using multiple antennas. Wirel. Pers. Commun. **6**, 311–335 (1998)
60. E.N. Gilbert, Random plane networks. SIAM J. **9**, 533–543 (1961)
61. P. Gupta, P.R. Kumar, The capacity of wireless networks. IEEE Trans. Inf. Theory **46**(2), 388–404 (2000)
62. J.F.C. Kingman, *Poisson Processes* (Oxford University Press, London, 1993). ISBN: 0198536933
63. M. Kang, M.-S. Alouini, G.E. Oien, How accurate are the Gaussian and gamma approximations to the outage capacity of MIMO channels? in *Proc. Baiona Workshop on Signal Processing in Communications*, 2003
64. A. Papoulis, S.U. Pillai, *Probability, Random Variables, and Stochastic Processes*, 4th edn. (McGraw-Hill, New York, 2002). ISBN:0071122567
65. T.K. Philips, S.S. Panwar, A.N. Tantawi, Connectivity properties of a packet radio network model. IEEE Trans. Inf. Theory **35**(5), 1044–1047 (1989)
66. S. Quintanilla, S. Torquato, R.M. Ziff, Efficient measurements of the percolation threshold for the fully penetrable disks. J. Phys. **13**(2), 722–731 (2003)
67. P. Santi, D.M. Blough, An evaluation of connectivity in mobile wireless ad hoc networks, in *Proc. IEEE DSN*, 2002
68. M.K. Simon, M.S. Alouini, *Digital Communication over Fading Channels: A Unified Approach to Performance Analysis* (Wiley, New York, 2000). ISBN:0471317799
69. H. Shin, J.H. Lee, Closed-form formulas for ergodic capacity of MIMO Rayleigh fading channels, in *Proc. IEEE ICC*, 2003
70. D. Watts, S. Strogatz, Collective dynamics of small-world networks. Nature (1998)
71. H. Yousefi'zadeh, H. Jafarkhani, M. Moshfeghi, Power optimization of wireless media systems with space-time block codes. IEEE Trans. Image Process. **13**(7), 873–884 (2004)
72. L. Zheng, H. Yousefi'zadeh, H. Jafarkhani, Resource allocation in fading wireless ad-hoc networks with temporally correlated loss, in *Proc. IEEE WCNC*, 2004

Index

S.J. Kazemitabar, *Coping with Interference in Wireless Networks*,
Signals and Communication Technology,
DOI 10.1007/978-90-481-9990-7, © Springer Science+Business Media B.V. 2011

MIX
Papier aus verantwortungsvollen Quellen
Paper from responsible sources
FSC® C105338

If you have any concerns about our products,
you can contact us on
ProductSafety@springernature.com

In case Publisher is established outside the EU,
the EU authorized representative is:
Springer Nature Customer Service Center GmbH
Europaplatz 3, 69115 Heidelberg, Germany

Printed by Libri Plureos GmbH
in Hamburg, Germany